ROYAL BOTANIC GARDENS, KEW

Kew Bulletin Additional Series II

A revision of the genus
HYPARRHENIA

W D CLAYTON

B SC, PH D, A R C S, F L S

LONDON
HER MAJESTY'S STATIONERY OFFICE
1969

SBN 11 240999 7

CONTENTS

A revision of the genus
Hyparrhenia

INTRODUCTION

Hyparrhenia is a genus of moderate size, with a distribution largely confined to the African continent. Its importance lies in the extent to which it dominates the herbaceous stratum of the African savannas. This is not the place to delve into ecological literature, but a glance at Rattray's (1960) map and commentary will make the point clear.

The most thorough taxonomic treatment of the genus is that of Stapf (1918). The general framework of his classification stands unscathed, but it was based, by modern standards, on very meagre herbarium collections which gave little inkling of the wide range of variation within each species. With the great increase in collections during the last 20 years or so it has become apparent that his discriminatory characters are often unsatisfactory, and in fact his work is now very difficult to use. Recent floras, whose accounts of *Hyparrhenia* are usually derived from Stapf's work, show evidence of this uncertainty. It is no exaggeration to say that the genus has been in a confused state, with many of the dominant savanna grasses habitually misidentified.

The taxonomic difficulties probably stem from the fact that polyploidy, apomixy and introgressive hybridization seem to be common phenomena in *Hyparrhenia*, and extensive cytogenetic research must be conducted before the genus is properly understood. Advances can also be expected to follow upon a better understanding of the ecological preferences of species, of which only the crude outlines are now apparent. However, detailed investigations of this kind take many years, and must be preceded by an examination of morphological variation and nomenclature conducted on classical lines, both to provide an interim classification, and to expose the problem areas to which the techniques of omega taxonomy might most profitably be directed. The present thesis, therefore, is primarily concerned with the gross morphology of specimens, as mainly represented in the Kew Herbarium. It has been accepted in partial fulfilment of requirements for the degree of Ph.D. in the University of London.

It became apparent early in the enquiry that the delimitation of the genus itself is by no means straightforward. It is therefore necessary to commence with a brief review of generic differentiation in the *Andropogoneae*, proceeding to a more detailed study of diagnostic criteria among the genera most closely allied to *Hyparrhenia*.

DELIMITATION OF THE GENUS

The tribe Andropogoneae

The *Andropogoneae* are a well-defined tribe best characterized by the occurrence of the spikelets in pairs, one sessile, the other pedicelled. It is inopportune to become embroiled in a detailed discussion of the tribal classification, but a brief outline is essential for a proper understanding of the evolution and systematic position of the genus *Hyparrhenia*.

The tribe is divided into 5 sub-tribes. Setting aside the enigmatic *Dimeriinae* (which does not have paired spikelets), three diverging lines can be recognized, corresponding to the sub-tribes *Saccharinae, Andropogoninae* and *Rottboelliinae*; *Ischaeminae* lies between the two latter, but seems to be more closely related to *Rottboelliinae*. The situation is visualized in figure 1.

Stapf (1918: 5–8) has divided the sub-tribes into groups of species (also Pilger, 1954; likewise without specification of rank). Since they are of slight taxonomic significance, there is little point in investing them now with the dignity and inflexibility of formal rank. However, applied informally, they are a useful aid to discussion. Employing these groups, morphological trends within the tribe as a whole may be summarized in the form of a key (not intended as a means of identification). The meaning of the descriptive terminology is discussed in some detail in the section on morphology (p. 9).

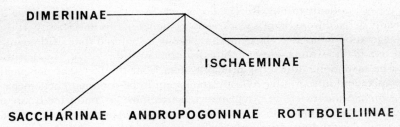

FIG. 1. Morphological affinities within *Andropogoneae*.

MORPHOLOGICAL TRENDS IN THE *Andropogoneae*

Spikelets solitary; rhachis tough . . . . DIMERIINAE

Spikelets paired:

Sessile and pedicelled spikelets both fertile; rhachis internodes and pedicels slender; callus blunt; lower palea small or absent; upper lemma usually bilobed; awn glabrous: (SACCHARINAE):

Inflorescence a large panicle; racemes tough or fragile; glumes usually thin and plumose; awns small . . . . . **Saccharastrae**

Inflorescence of solitary or digitate racemes; racemes fragile; glumes indurated, and hairy but not plumose; awns usually well developed **Eulaliastrae**

Sessile spikelet alone fertile, the pedicelled male or more or less suppressed; racemes fragile:

Rhachis internodes and pedicels thickened; racemes 1-several, rarely paniculate; lower glume often muricate or otherwise armoured; callus blunt; lower palea usually present; upper lemma bilobed; awn glabrous:

Sessile spikelet awned; internode and pedicel forming a stout V-shaped frame behind sessile spikelet: (ISCHAEMINAE):
Racemes of many spikelets, espathate   .   .   .   **Ischaemastrae**

Racemes reduced to 1 sessile spikelet and 2 involucral pedicels, the whole enclosed in a spatheole   .   .   .   .   **Apludastrae**

Sessile spikelet awnless, glabrous; lower glume, internode and pedicel (the two latter sometimes fused) together forming a 3-sided or barrel-shaped box enclosing sessile spikelet; pedicelled spikelet commonly suppressed: (ROTTBOELLIINAE):
Racemes digitate or paniculate, espathate   .   .   **Vossiastrae**

Racemes solitary, often spathate   .   .   .   **Rottboelliastrae**

Rhachis internodes and pedicels slender; lower palea small or absent: (ANDROPOGONINAE):
Inflorescence a panicle   .   .   .   .   .   .   **Sorghastrae**

Inflorescence usually reduced to solitary or paired racemes:
Lower glume of sessile spikelet 2-keeled; callus obtuse, inserted; awn glabrous; internodes and pedicels sometimes thickened:
Upper lemma entire; inflorescence sometimes a panicle, espathate
Bothriochloastrae

Upper lemma bilobed; inflorescence never a panicle, sometimes digitate, often spathate; upper glume occasionally awned
Andropogonastrae

Lower glume of sessile spikelet rounded; callus pungent, oblique; awn hairy; inflorescence usually spathate:
Pedicelled spikelet without a callus; racemes paired; upper lemma bilobed   .   .   .   .   .   .   **Hyparrheniastrae**

Pedicelled spikelet with a callus; racemes usually solitary:
Callus of pedicelled spikelet short; homogamous spikelets 0 to many; upper lemma bilobed; upper glume awned
Anadelphiastrae

Callus of pedicelled spikelet long; homogamous spikelets several to many; upper lemma entire; upper glume awnless:
Homogamous spikelets forming part of a linear raceme
Heteropogonastrae

Homogamous spikelets forming an involucre around a reduced raceme   .   .   .   .   .   **Themedastrae**

The topic will be re-opened in subsequent sections; for the present it will suffice to concentrate upon the sub-tribe *Andropogoninae*. The morphological relationships between the species groups of this sub-tribe will become clearer if they are visualized as forming a linear sequence (Fig. 2) which progresses from simple to highly differentiated structures. Although doubtless reflecting, in some measure, the sequence in which the various characters have evolved, the simple linear series is certainly an oversimplification. The most obvious of the possible modifications is to divide it into two parallel lines, as is commonly done in conventional keys, according to whether the upper lemma is entire or bilobed. In view of the remarkable similarity between *Euclasta* Franch. and *Pseudodichanthium* Bor (*Bothriochloastrae*) on the one hand with *Agenium* Nees and *Heteropogon* Pers. (*Heteropogonastrae*) on the other, this dichotomy may well represent a better approximation to phylogeny (see also p. 29).

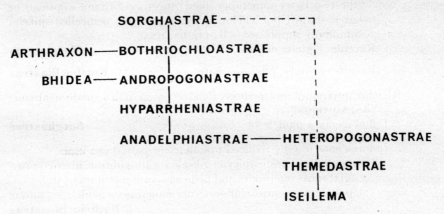

FIG. 2. Morphological affinities within *Andropogoninae*.

To attempt a phylogenetic arrangement would involve endless speculative elaboration. For the present purpose the linear sequence is sufficient, for it makes clear the major morphological trends within the *Andropogoninae* upon which the generic classification is based. The trends are listed below:

1. Inflorescence: simple, espathate → compound, spathate.

2. Racemes: paniculate → digitate → paired → solitary.

3. Racemes comprising: many fertile spikelets → 1 fertile spikelet with attendants.

4. Pedicels and rhachis internodes: slender → thickened.

5. Homogamous spikelets: none → several → involucral.

6. Callus of sessile spikelet: obtuse, inserted → pungent, oblique.

7. Lower glume of sessile spikelets: dorsally flattened $\Big\langle \begin{array}{l} \nearrow \text{2-keeled} \\ \searrow \text{semicylindrical.} \end{array}$

8. Upper glume of sessile spikelet: awnless $\rightarrow$ awned.

9. Awn of fertile floret: glabrous $\rightarrow$ hirsute.

10. Pedicelled spikelet: short callus $\rightarrow$ long callus $\rightarrow$ fertile.

It may be noted that the 2-keeled lower glume, thickened pedicels and awned upper glume reach their zenith part way along the proposed linear series, and constitute a further source of complications to any phylogenetic scheme that might be attempted.

The allied genera

In order to arrive at a satisfactory definition of *Hyparrhenia*, it is now necessary to examine in more detail those genera most closely related to it. They may be defined as those members of the *Andropogoneae* having solitary, paired or digitate racemes, the fertile lemma awned from the sinus of its 2-lobed tip, and the lower floret of the sessile spikelet reduced to a barren lemma. They are the genera falling into Stapf's groups *Andropogonastrae*, *Hyparrheniastrae* and *Anadelphiastrae*.

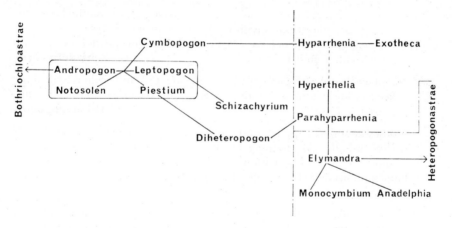

FIG. 3. Morphological affinities of *Hyparrhenia* and its allies.

Although the designation of genera within the *Andropogoneae* has often been regarded as peculiarly difficult, it is doubtful whether intergeneric boundaries are more obscure than in many temperate tribes. Much of the difficulty probably lies in the fact that we have not become conditioned by familiarity to accept certain inconspicuous characters as the final arbiters of generic limits, as we have in the case of temperate grasses. In fact a number of excellent characters may be found, the most important being those discussed in the previous paragraphs. Their application is best demonstrated by means of a diagnostic key, which is followed by brief notes on each genus, whose interrelationships are shown diagrammatically in figure 3.

KEY TO GENERA ALLIED TO HYPARRHENIA

Callus of sessile spikelet inserted in the cupuliform or crateriform apex of the internode; lower glume of sessile spikelet 2-keeled, flat or depressed between the keels (except *Schizachyrium*); homogamous spikelets inconspicuous (except *Diheteropogon*) (*Andropogonastrae*):

Callus of sessile spikelets obtuse, usually very short; awn glabrous to scaberulous:

Racemes paired or digitate, rarely solitary but then nerveless between the keels of the lower glume; lower glume of sessile spikelet flat or concave on the back, the keels lateral or dorsal:

Racemes not deflexed, borne upon unequal more or less terete raceme-bases, usually exserted; homogamous pairs absent, or scarcely different from heterogamous pairs; pedicels and rhachis internodes linear to clavate or swollen; leaves not aromatic
1. **Andropogon**

Racemes deflexed at maturity, borne upon subequal flattened raceme-bases, 1–2·5 mm. long, seldom exceeding the spatheole; homogamous pairs 1 at the base of the lower raceme, the pedicel often swollen; pedicels and rhachis internodes linear; leaves aromatic; panicle dense, decompound . . . .   2. **Cymbopogon**

Racemes solitary; lower glume of sessile spikelet convex on the back, the keels lateral or frontal with several intercarinal nerves; pedicels and rhachis internodes clavate to linear, the apex deeply hollowed into a cup with a fimbriate rim . . . . .   3. **Schizachyrium**

Callus of sessile spikelets pungent, 1–5 mm. long; awn puberulous to hirsute; keels of lower glume rounded with 2–4 nerves in each keel; racemes paired . . . .   4. **Diheteropogon**

Callus of sessile spikelet applied obliquely to the apex of the internode with its tip free, usually acute to pungent; lower glume of sessile spikelet convexly rounded on the back without keels (rarely with a median groove); homogamous spikelets often present; pedicels and rhachis internodes linear:

Pedicelled spikelet without an appreciable callus, and upper glume of sessile spikelet awnless (sometimes possessing these features, but then lower glume with a median groove); fertile lemma usually minutely bidentate, its awn pubescent to hirtellous (*Hyparrheniastrae*):

Lower glume of sessile spikelet with a median longitudinal groove; upper glume acute to apiculate or shortly awned:

Raceme-base without a distinct appendage; racemes 9–14-awned per pair; lower glume of sessile spikelet without a conspicuous herbaceous tip . . . .   5. **Parahyparrhenia**

Raceme-base produced at the tip into a long scarious appendage; racemes (1–)2(–10)-awned per pair; lower glume of sessile spikelet with a distinct herbaceous or membranous tip   6. **Hyperthelia**

Lower glume of sessile spikelet rounded on the back, or rarely with 2 or more shallow striations; upper glume obtuse to acute or mucronate: Upper raceme-base up to 10 mm. long, but usually much shorter

7. **Hyparrhenia**

Upper raceme-base 15–25 mm. long; homogamous pairs 2 at the base of each raceme, forming an involucre. . . 8. **Exotheca**

Pedicelled spikelet prolonged at the base into a callus 0·5–3 mm. long; upper glume of sessile spikelet awned; lower glume without a median groove (except *Anadelphia scyphofera*); fertile lemma usually bifid for $\frac{1}{4}$–$\frac{1}{2}$ its length, the awn shortly pubescent to glabrescent (*Anadelphiastrae*):

Racemes typically paired, with 1–10 homogamous pairs at the base of the lower or both, exserted from the narrow spatheole

9. **Elymandra**

Racemes solitary, without homogamous pairs:
Spatheoles cymbiform, coloured and enclosing the racemes; racemes dense, with the spikelets concealing the short internodes; pedicelled spikelets broadly lanceolate, subacute, hairy or merely spinose-ciliate on the margins; callus of sessile spikelet obtuse

10. **Monocymbium**

Spatheoles linear to narrowly lanceolate, usually greenish, embracing the racemes or not; racemes loose, with few spikelets and long internodes visible between them; pedicelled spikelets linear-lanceolate, acuminate, nearly always glabrous; callus of sessile spikelet usually acute to pungent . . . . 11. **Anadelphia**

1. **Andropogon** Linn. A rather heterogeneous pan-tropical genus, discussed by Clayton (1967). It is remarkable for the hollow back and twin keels of the lower glume of the sessile spikelet, seen at their most extreme form in sect. *Piestium* Stapf. These, together with the glabrous awn and inserted callus, distinguish the genus from *Hyparrhenia*, although there is some difficulty with a few species in which the diagnostic characters are not well developed. One such is *A. pusillus*, transferred from *Hyparrhenia* for reasons which are discussed elsewhere (Clayton, 1965). Another, *A. wombaliensis*, has here been transferred to *Hyparrhenia*.

2. **Cymbopogon** Spreng. A homogeneous genus of the Old World tropics whose relationships have been commented upon earlier (Clayton, 1965). Like its ally *Andropogon*, it is distinguished from *Hyparrhenia* by a 2-keeled lower glume and obtuse callus. However, it has short flattened deflexed raceme-bases such as occur in some sections of *Hyparrhenia*, and the possession of so important a feature in common suggests that the two genera are related. Indeed the resemblance between *H. glabriuscula* and *Cymbopogon* is quite remarkable.

3. **Schizachyrium** Nees. The delimitation of this genus and the characters distinguishing it from *Andropogon* have been discussed by Clayton (1964). Owing to its solitary racemes and crateriform internode tip it cannot be confused with *Hyparrhenia*.

4. **Diheteropogon** Stapf. An African genus closely related to *Andropogon* (*A. ivorensis* Adjanohoun & Clayton (1963) is almost exactly intermediate between the two), but differing in the rounded several-nerved keels separated by a narrow groove, hairy awn, and pungent sessile spikelet callus (Clayton, 1966a). The latter characters ally it to the *Hyparrheniastrae*, but with the important distinction that the callus is thrust deeply down within the internode.

5. **Parahyparrhenia** A. Camus. A small West African genus (see Clayton, 1966e) similar in structure to both *Diheteropogon* and *Elymandra*. It has the grooved lower glume of *Diheteropogon*, but its callus is free from the internode. On the other hand it lacks the conspicuous pedicelled spikelet callus found in *Elymandra* (the latter does not have a grooved glume, although the character crops up occasionally in allied genera such as *Anadelphia* and *Heteropogon*). It thus falls naturally between the *Andropogonastrae* and *Anadelphiastrae*, but has at least the artificial characters of the *Hyparrheniastrae*, being separated from *Hyparrhenia* itself chiefly by the grooved glume.

6. **Hyperthelia** W. D. Clayton. Primarily an African genus created from *Hyparrhenia* by the transfer of four species with grooved lower glumes and a tendency to awned upper glumes, the reasons being discussed in more detail by Clayton (1966e). It is clearly related to *Parahyparrhenia*, and also seems to link that genus with *Hyparrhenia* Sect. *Hyparrhenia*, but the reality of this linkage is open to question and will be examined later.

7. **Hyparrhenia** Fourn. (See below.)

8. **Exotheca** Anderss. A satellite genus allied to *Hyparrhenia* sect. *Apogonia*, and reviewed by Clayton (1966e). It is separated more because intermediates are lacking, than because of the novelty of its characters.

9. **Elymandra** Stapf. An African genus distinguished from *Hyparrhenia* by its awned upper glume, and pedicelled spikelets supported on a callus; the general facies and colouring are also characteristic. These features are also found in *Hyparrhenia* sect. *Archaelymandrae* Jac.-Fel. and series. *Grallatae* Stapf, whose species are clearly misplaced in that genus and have been transferred to *Elymandra* (Clayton, 1966d).

10. **Monocymbium** Stapf. An African genus reviewed by Jacques-Félix (1950). The solitary racemes and awned upper glume serve to separate it from *Hyparrhenia*.

11. **Anadelphia** Hack. An African genus revised by Clayton (1966c). With its solitary racemes and well-developed pedicelled spikelet callus it cannot be confused with *Hyparrhenia*.

The genus Hyparrhenia

Having established the criteria for separating *Hyparrhenia* from its allies, it remains to enquire whether the remaining species are sufficiently coherent to be treated as a single genus.

Stapf (1918) grouped the species of *Hyparrhenia* as follows:

Sect. *Eu-Hyparrhenia* (=sect. *Polydistachyophorum*)
    Series 1. *Rufae*
         2. *Hirtae*
         3. *Grallatae*   (Transferred to *Elymandra*)
         4. *Filipendulae*
Sect. *Ruprechtia*
         5. *Ruprechtiae*   (Transferred to *Hyperthelia*)
Sect. *Pogonopodia* (incl. sect. *Hyparrhenia*)
         6. *Cymbariae*
         7. *Bracteatae*
Sect. *Apogonia*
         8. *Diplandrae*
         9. *Involucratae*
        10. *Cornucopiae*   (Transferred to *Hyperthelia*)
Sect. *Dibarathria*
        11. *Pusillae*   (Transferred to *Andropogon*)

Four of these series have been transferred to other genera for reasons previously discussed. Of the sections that remain *Polydistachyophorum* is generally considered to be the most distinctive owing to its terete raceme-base; indeed Jacques-Félix (1962) has intimated that generic rank might be appropriate. However, as will be seen later, it is by no means easy to distinguish between species in sects. *Polydistachyophorum* and *Pogonopodia* (consider *H. hirta* and *H. dregeana*, or *H. gazensis* and *H. pilgerana*), while the species in the new sect. *Strongylopodia* provide a further stumbling block to a generic distinction based on the shape of the raceme-base. The inevitable conclusion is that this character provides an unsatisfactory basis for the separation of a new genus.

The present revision indicates that *Apogonia* is the section most worthy of elevation, but there are obvious difficulties over *H. gossweileri* and sect. *Arrhenopogonia* and I do not believe that generic rank is appropriate.

MORPHOLOGY

Vegetative characters

The species of *Hyparrhenia* are tall (1–3 m.) erect grasses of the open savanna, including 14 annuals and 38 perennials, but in 3 species the culm is slender and straggling. The perennating organ is commonly referred to as a short rhizome, but rootstock might be a better word as all the species form compact clumps. Stilt roots are often present, and may be so well developed that the slender connection to the rhizome is often overlooked unless the plant is carefully collected.

The leaf-blades are long and linear (except in the straggling species) and usually of the somewhat glaucous green colour typical of savanna vegetation. The short scarious ligule is unremarkable except in sect. *Hyparrhenia* which has a strong tendency to long ligules associated with sheath auricles and a falsely petiolate leaf-base. The sheath is of interest only in those few species whose basal sheaths bear a pubescent or tomentose indumentum. Charred sheaths left by the bush fires of the previous year are a regular feature in perennials.

The first seedling leaf is usually narrowly lanceolate and ascending.

In general the vegetative features are very uniform, and provide few useful taxonomic characters.

Anatomy

The abaxial leaf epidermis may be described as follows: silica bodies cross to dumb-bell shaped, occasionally nodular; micro-hairs 2-celled, the distal cell tapering; stomatal subsidiary cells triangular or dome-shaped.

Transverse sections of the leaf show radiate chlorenchyma, single bundle sheaths and angular small vascular bundles.

The anatomy is of the typical panicoid type, and shows little difference in the various published descriptions: *H. hirta* (Prat, 1932); *H. filipendula* (Vickery, 1935); *H. cyanescens* (Prat, 1937); *H. cymbaria, H. diplandra, H. filipendula, H. hirta, H. lintonii, H. rufa* (Stewart, 1965). Nor does the leaf anatomy of *Hyparrhenia* appear to differ significantly from any of the allied genera described by Metcalfe (1960). It is clear that this is not a promising field in which to seek taxonomic distinctions, and it has not been pursued.

Reichwaldt (1945) has described the anatomy of the glumes in *H. nyassae*, and Tran (1965) that of the awn in *H. subplumosa*. Krupko (1956) found that the primary root of *H. aucta* (=*H. dregeana*) was nearly always hexarch, whereas that of *H. hirta* was usually pentarch.

Panicle

The inflorescence is made up of paired racemes borne in a copiously branched false panicle, false because its branches are subtended by modified leaves or *spathes*; it is in fact a specialized branch system, only the ultimate units being a true inflorescence. The spathes are at first leaf-like, but with successive degrees of branching they become increasingly modified, usually by reduction of the limb, inflation of the sheath and adoption of a reddish colour. The ultimate, and most modified, of them is termed a *spatheole*.

The structure of the false panicle has been described by Stapf (1918: 209). One or two of the lowest branches of the panicle may start with an elongated internode and thereafter behave like the primary axis. With this exception each node of the primary axis gives rise, within the axil of the subtending spathe, to a very short branch or *sympodium* which undergoes a rapid sequence of cymose branching. The very short internodes of the sympodium commence with an adaxial *prophyll*. This is a 2-keeled, linear, membranous structure, seldom exceeding the spathe in length, concave on the back, and enfolding the younger parts of the sympodium. The prophylls are very similar in each species, and seem to be of little taxonomic value. At each of the nodes an adaxial branch is thrown off in the axil of the prophyll and the sympodium itself is continued alternately slightly left or right of the centre line (Fig. 4).

Branches from the sympodium commence with an elongated internode and are termed *rays*; those from each node of the primary axis together form a *tier*. The rays may be *simple*, consisting of a single internode terminating in a spatheole; or they may be *compound*, with several nodes each giving rise to a secondary tier. By repeating this process a most complicated structure may be built up (Fig. 4).

The appearance of the panicle is often a useful aid to recognition, but a detailed analysis of its structure is not usually helpful. Such an analysis is tedious to perform, difficult to express concisely, and displays so much variation that it has little meaning when applied to individual plants. For this reason the formal description of the panicle is given only in general terms. It has been shown that in several species a 'witches broom' clumping of the false panicle is induced by nematode infestation (Corbett, 1966).

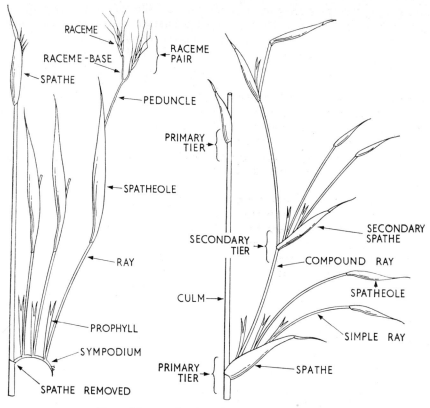

Fig. 4. Diagram showing structure of false panicle.

Spatheole and peduncle

The shape, length and colouring of the spatheole are among the most important taxonomic characters. The spatheoles resemble inflated leaf-sheaths without a limb, and vary from narrowly ovate boat-shaped structures to linear organs tightly rolled around the peduncle at maturity. They are often brightly coloured, and are usually glabrous though sometimes a few hairy specimens are found; the latter does not seem to be of much taxonomic significance.

The peduncle is filiform, and its length relative to the spatheole varies greatly. As a general rule it is short when the racemes deflex, and longer than the spatheole when they do not, but there are many exceptions. It is commonly adorned with long spreading hairs in its upper part, but glabrous specimens are to be found in most species. Thus, although a most conspicuous feature, this character is seldom of any taxonomic value. The articulation between peduncle and raceme-base may be decorated by a ring of hairs borne on the foot of the raceme-base.

Raceme-base

Each raceme is supported upon a short stalk called a *raceme-base*, the two bases fusing together just above the point at which they are articulated with the peduncle (but in odd specimens of species such as *H. rufa* they may be connate for some distance above the foot). One of them (conventionally termed the *lower*) is very short and the raceme it bears is subsessile, but the other (*the upper*) is more variable in length. The morphological nature of the raceme-base may be ascertained from species in neighbouring genera with 3 or more racemes, such as *Andropogon lateralis*; and from occasional abnormal specimens of *Hyparrhenia* with branched racemes. It is then seen that branching is accomplished by the production of two rhachis internodes side by side in an otherwise normal raceme-segment. The raceme-bases are clearly homologous structures and, with their accrescent pair of homogamous spikelets, can be regarded as parts of a modified raceme-segment with twin internodes. The raceme-bases (and homogamous spikelets) remain attached to the plant for some time after the rest of the raceme has disintegrated, and specimens in this condition fre quently puzzle the unsuspecting beginner. Eventually the raceme-bases also fall.

Two main types of raceme-base may be recognized; that in which the upper base is terete and much longer than the lower, and that in which it is strongly flattened and scarcely exceeds the lower. Bases of the latter class nearly always deflex at maturity, but the phenomenon is not so firmly established in the former group. *H. madaropoda* is interesting in that deflexion of the racemes is achieved by the peduncle rather than the raceme-bases.

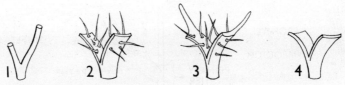

Fig. 5. Diagrammatic representation of raceme-base types (homogamous spikelets omitted). **1**, sect. *Polydistachyophorum*; **2**, sect. *Pogonopodia*; **3**, sect. *Hyparrhenia*; **4**, sect. *Apogonia*.

The raceme-bases are commonly pubescent in the fork and glabrous above, or occasionally the upper raceme-base may be sparsely pilose. In two sections however the raceme-bases are obscured in coarse stiff glassy bristles, which are seldom difficult to distinguish from the softer and sparser hairs sometimes found in other sections.

The tip of the raceme-base may be truncate, oblique, or drawn out into a scarious appendage from its upper edge. The presence of such an appendage is used to distinguish sect. *Hyparrhenia*, but its use presents some difficulties for the transition from oblique or scariously rimmed to obviously appendaged is very gradual, and its correlation with the other characteristics of the section is not absolute. Since the raceme-bases are concealed by stiff hairs in this section, the appendage, which is not over 4 mm. long, must be searched for with care. Note also that the appendaged tip of the raceme-base occurs above the point of insertion of the homogamous spikelets if these are present.

The raceme-bases are of the greatest taxonomic importance, for they provide the characters by which the sections are primarily delimited (Fig. 5). In general the different types are quite easy to recognize, but there are one or two anomalies which can be confusing. The species most likely to mislead are:

    H. glabriuscula: raceme-base ambiguous
    H. gazensis: upper base densely pilose
    H. andongensis. bases appendaged
    H. cyanescens: bases often only sparsely hairy

H. papillipes: bases appendaged
H. niariensis var. *macrarrhena:* appendage absent
H. madaropoda: upper base long, subterete; appendage small
H. newtonii: upper base often rather long and subterete
H. gossweileri: bases unequal

Racemes

The racemes occur in pairs, but in *H. mobukensis* solitary racemes are common. Rarely, and obviously aberrantly, more than two may be found. The racemes are comparatively short and contain relatively few spikelets, sometimes only a single triad. The number of spikelets present is of great value in separating species, but difficult to count as mature racemes have often started to shatter. The easiest method is to count the number of awns protruding from an unopened spatheole; the number of spikelets is therefore given as the number of awns per raceme-pair.

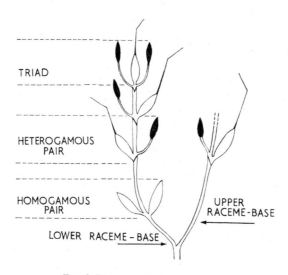

TRIAD

HETEROGAMOUS
PAIR

HOMOGAMOUS
PAIR

UPPER
RACEME-BASE

LOWER RACEME-BASE

Fig. 6. Diagram of raceme structure.

Each fertile (sessile) spikelet is accompanied by a pedicelled spikelet, but the terminal sessile spikelet has two pedicelled attendants. One or two of the lower pairs often differ from the rest, and are termed *homogamous*. There is always a pair of homogamous spikelets at the base of the lower raceme. The presence of further pairs is of taxonomic importance, but the character is not always easy to use for the number of additional pairs is seldom absolutely constant; indeed in some species it may be rather variable (see Fig. 6).

Homogamous spikelets

Both members of the pair are alike, more or less sessile, and fused to the adjacent rhachis internode (or raceme-base). In general they resemble the pedicelled spikelets, being male or barren and somewhat larger than the sessile spikelets, but they are always awnless. The internodes are about the same length as those in the rest of the raceme, except in sect. *Apogonia* where the two pairs are very close together and form a sort of involucre embracing the lower part of the raceme.

Sessile spikelets

There is a bearded prolongation at the base of the spikelet, conveniently referred to as the callus of the sessile spikelet. However, since it occurs below the
point of insertion of the spikelet on the rhachis and forms part of the articulation between adjacent internodes, it is probably more correct to regard it as a
nodal structure. The presence of an obtuse callus in a few species is a useful
character, somewhat mitigated by a tendency to intergrade with the usual
pointed shape, and by the difficulty of measuring its length accurately owing
to uncertainty about its upper limit. The callus is always applied obliquely to
the top of the internode with the tip free and often prominently projecting
(occasional aberrant specimens may have the rim of the raceme-base lapping
over the callus tip).

The lower glume is rounded on the back, rarely striate or slightly hollowed.
It is unobtrusively sub-coriaceous, but the glossy texture in *H. rufa* is distinctive.
The shape is useful only in general terms for it changes to some extent as the
fruit matures, but the indumentum is of taxonomic interest. The tip is usually
emarginate or bimucronate to fit snugly about the base of the awn. The upper
glume is concealed between the pedicel and internode, and seems to be of little
taxonomic value.

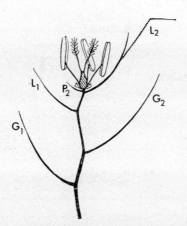

Fig. 7. Diagram of spikelet structure. **G** = glume, **L** = lemma, **P** = palea

Within the glumes are two lemmas. The lower is a featureless hyaline scale
without floral parts. The upper is a hyaline scale with two short lobes at the
apex between which the awn arises, but by the time of maturity it has become
little more than a flattening at the base of the awn. Between the two lemmas can
sometimes be found a tiny ovate scale up to 0·5 mm. long. This can only represent a palea, presumably of the upper floret, and confirms the interpretation
of the spikelet as a two-flowered structure, only a lemma remaining to indicate
the presence of a lower floret (Fig. 7). The upper floret is hermaphrodite
with the usual 2 small lodicules, 3 stamens and 2 stigmas. The caryopsis has an
embryo about one third its length, and a basal hilum. The embryo is of the
usual panicoid type with vascular internode, no epiblast, a scutellum cleft and
a rolled first leaf (Fig. 8), corresponding to Reeder's (1957) formula P-PP.
See also Kinges (1961), who has examined the embryos of *H. hirta* and
'*H. buchananii*'.

The awn is usually stout and conspicuous, geniculate, with a hairy twisted column and straight tapering limb. Most sections contain one or two species with unusually long column hairs, but they are characteristic of the *Filipendula*-group. In two species the awn is absent, a condition which is easily confused with abortion of the awns due to smut attack, and it is essential to dissect out the upper lemma before a naturally awnless state is accepted.

Spikelet size increases in the order *Pogonopodia–Polydistachyphorum–Arrheno-pogonia–Apogonia–Hyparrhenia*. As is to be expected spikelet length is correlated with callus and awn length, and inversely related to the number of spikelets per raceme. However such statements are gross generalizations of little taxonomic value.

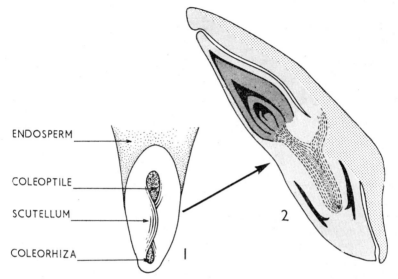

ENDOSPERM

COLEOPTILE

SCUTELLUM

COLEORHIZA

2

1

FIG. 8. *H. involucrata*. Diagrams to show structure of embryo. **1,** surface view, pericarp removed; **2,** median logitudinal section.

Pedicelled spikelets

The pedicelled spikelets are usually longer than the sessile. The extension of the lower glume into an awn in some species is a useful, but often imprecise, aid to diagnosis. Within the glumes are two hyaline awnless lemmas, the upper subtending 3 stamens. The basal callus is short, rounded and obscure. Occasionally it is more definite, but is then broad and clearly part of the spikelet, rather than apparently part of the pedicel as in the *Anadelphiastrae*. The presence of a little tooth at the tip of the pedicel is a most helpful character for distinguishing certain species.

CYTOLOGY

Cytological knowledge of *Hyparrhenia* is meagre. Several new chromosome counts are reported here, but this line of enquiry was limited in scope. The counts were made on Feulgen-stained root-tip squashes, after pre-treatment for 4 hours in alpha-bromo-naphthalene or overnight at $1\,°C$ in distilled water. Root maceration was aided by a treatment in 3% pectinase solution. I should like to express my thanks to Mrs. M. Haylock and other members of the Jodrell Laboratory staff who made the slides, and to the many field collectors who provided seed.

The chromosome numbers counted or reported in the literature are summarized in Table 1. The numbers are not very informative, but it is clear that 10 is the major basic number, with 9 and 15 as other possibilities. It can be seen that polyploidy has been an active factor in speciation within the genus. Brown & Emery (1957a) have found persistent nucleoli in *H. hirta*, as is to be expected in a panicoid genus.

TABLE 1. Chromosome numbers recorded in *Hyparrhenia* (full references for each species are given below under Taxonomy)

Species	2n	Species	2n
H. smithiana	c. 20, 40	H. cyanescens	c. 36
H. nyassae	20, 40	H. dregeana	20
H. rufa	30, 36, 40	H. niariensis	18
H. gazensis	30	H. welwitschii	40
H. hirta	30, 40, 44, 45, 46	H. newtonii	40
H. filipendula		H. arrhenobasis	22
var. pilosa	40	H. involucrata	
H. cymbaria	20, 30	var. breviseta	40
H. variabilis	20	H. mutica	60
H. formosa	30	H. subplumosa	40
H. schimperi	40	H. diplandra	20

Brown & Emery (1957b, 1958) have also reported the presence of 4-nucleate embryo sacs in *H. rufa* and *H. hirta*, from which they infer that these species are apomictic. They have shown that this phenomenon is common in panicoid grasses, and it is quite likely to occur in other species of the genus.

Mehra & Anderson (1960) have analysed the morphological variation between *H. cymbaria* and '*H. papillipies*' (probably *H. dregeana*) growing together along pathsides in Ethiopia, and shown that it bears all the hallmarks of introgressive hybridization. Similar evidence appears repeatedly in the taxonomic part of the present paper, and it is scarcely possible to doubt that the phenomenon is of widespread occurrence in the genus.

Thus although cytological knowledge is fragmentary, it seems clear that three classical elements of speciation—polyploidy, apomixis and hybridization—are represented in good measure. The situation recalls the evolutionary model discussed by Zohary & Feldman (1962) and Stebbins (1956), and gives fair warning that boundaries between species are likely to be confused and indistinct, and that the taxonomy of the genus will not be easy.

ECOLOGY

The savanna environment

Hyparrhenia is pre-eminently a species of the African savannas. Indeed it is a fair approximation to observe that this habitat is characteristic of the *Andropogoneae* as a whole, for even in tropical Asia, where extensive savannas do not occur, the environment favoured by the *Andropogoneae* is very similar.

The African land surface is typically a flat pediplain, surmounted here and there by residual inselbergs. The step between successive pediplains, the most extensive of which lies at about 1300 m., sometimes takes the form of fine escarpments, but is often merely a zone of dissected country (Clayton, 1958b). The soils vary enormously, but they are usually more or less sandy, and the profile often shows a horizon of sesquioxide enrichment at a depth of a metre or so. Where it has been exposed by erosion this horizon may harden into a pan of ironstone inhospitable to plant life, which is particularly impressive when forming protective caps on flat-topped hills. Beyond the larger rivers there are few streams, drainage being effected by almost imperceptible hollows which form narrow treeless grassy tracts of grey sand or black montmorillonitic clay winding across the pediplain, and along which the water gently seeps.

The rainfall, of about 65–125 cm. per year, is quite sufficient to support a continuous vegetation cover of grass below and openly spaced trees above. However the bulk of it falls in only 7 months of the year, so that the plants must endure a long dry season. The alternation of drought and waterlogging is extreme on the clayey soils of drainage hollows. Heavy thunderstorms 2 or 3 weeks in advance of the steady rains are a characteristic feature.

The savanna is loosely described as a climax formation. In fact it is a fire disclimax because, for a variety of reasons, the vegetation is fired nearly every year; indeed so prevalent is the burning that the nature of the climatic climax—probably some form of woodland—has never been satisfactorily established. The frequency of burning has certainly increased in recent times, but there is evidence that it has always been an important factor of the savanna environment. Lightning has been observed to start bush fires, and charcoal fragments are a feature of certain soil profiles. More telling however is the remarkable ability of the thick-barked savanna trees to withstand these annual fires, which quickly kill forest trees exposed to the same conditions (Clayton, 1958a & 1961). It seems certain that savanna burning is of sufficient antiquity to have been a potent factor in natural selection.

Most of the savanna is subject to cultivation, but the fertility of the soil is low. Under the prevalent system of subsistence agriculture fertilizers are uneconomic, and a short spell of cultivation must be followed by a long period of recuperative tumbledown fallow (Clayton, 1959 & 1963). The fallow is initiated by ruderal species of the *Aristideae* and *Eragrostideae*, and the ground layer sere proceeds to a bunch-grass dis-climax of *Andropogoneae*. *Hyparrhenia* is a typical constituent of the dis-climax, and its establishment is commonly taken as a sign that the land is once again fit for cultivation. In short the environment in which the genus thrives presents ample opportunities for what Anderson (1948 & 1949) has called 'hybridization of the habitat'. Species which cannot normally compete successfully in one another's territory owing to slightly different ecological preferences, are able to grow together in such a habitat. Frequent cytological introgession is to be expected in these circumstances, with consequent blurring of taxonomic boundaries.

Habit and habitat

Hyparrhenia is a genus which has adapted itself to the savanna environment, ranging widely within the savanna but little outside it. Table 2 (p. 18) indicates very roughly the number of species in different habitats.

As might be expected the vegetative organs are relatively uniform, there being little opportunity for other than tall erect grasses in such an environment.

In fact there is little plasticity in this respect among the *Andropogoneae* as a whole. There is however a tendency for annuals to be rather common in sect. *Hyparrhenia*, and rare in sect. *Pogonopodia*. If, as will be suggested later, the former is considerably older than the latter, some support can be found for Stebbins's (1957) proposition that annuals tend to evolve from perennials, a process which has developed in sect. *Hyparrhenia*, but not yet in the younger section.

TABLE 2. Number of *Hyparrhenia* species in different habitats

Habitat	Number of species
Savanna grassland	21
Wet grassland	13
Open sites with dry soils	6
Pathsides and old farmland	9

The basal felt of *H. smithiana* and *H. nyassae* is interesting, but of doubtful function. Verboom (personal communication) has observed that it is often soaked with dew in the morning, and may be concerned with conserving water from this source. It is also very prominent in the dry season, and may serve as additional protection, for the dormant buds, from fire or insolation.

In passing it may be noted that long wave radiation is reflected by white, yellow or reddish-brown, and absorbed by black; while short wave lengths (ultra-violet) are reflected by yellow, reddish-brown or black (Bonsma, 1949). The characteristic colouring of the racemes in many species—yellowish hairs shielding dark coloured spikelets—may thus be seen as a protective adaptation against insolation.

Boughey *et al.* (1964) and Munro (1966) have recently found that some species of *Hyparrhenia* secrete a toxin which suppresses the growth of nitrifying bacteria. The intriguing ecological implications of this factor on the nitrogen deficient soils of the African savanna have not yet been followed up.

Flowering behaviour

Few observations have been reported on the flowering behaviour of *Hyparrhenia*. Agreda & Cuany (1962) found that *H. rufa* behaved as a short-day plant with a critical day-length of 12 hr. 15 min., with little difference between plants procured from different latitudes and altitudes. Mes (1952) found that *H. hirta* was also a short-day plant with a similar critical day-length.

FORM, FUNCTION AND EVOLUTION

The raceme-segment

The most characteristic feature of the *Andropogoneae* is the fragile raceme disarticulating between each pair of spikelets. The disseminule is thus a composite unit made up of sessile spikelet, rhachis internode, pedicel and pedicelled spikelet. Many important characters used for generic discrimination can be attributed to the exploitation of the possibilities inherent in this unit. Unfortunately it is not at all clear how this feature originated, or what advantage its primitive form could confer. Presumably it offers some biological advantage, for paired spikelets (but without an articulated rhachis) are found in several genera of the *Paniceae*. It is tempting, but probably fanciful, to regard the sub-tribe *Dimeriinae* (*Saccharum*-like spikelets on a *Brachiaria*-like rhachis) as intermediate between the two tribes.

The *Saccharastrae* are customarily considered the most primitive group in the *Andropogoneae*, for it has many of the features which one would expect to find in an archetype: the large panicle, similar spikelets, thin glumes, poorly developed awns, and slender rhachis which often remains intact at maturity, the spikelets breaking off below the glumes (Fig. 9/A). However, the plumose spikelets are efficient agents of wind dispersal, and the other characters, although possibly of a relict nature, can be seen as complementary adaptations for the same purpose. Be that as it may, in both groups of the *Saccharinae* the articulated rhachis is firmly established and they give no clue to its origin.

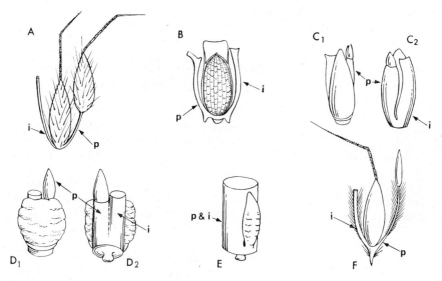

Fig. 9. Raceme segments in *Andropogoneae*. **A**, *Saccharum*; **B**, *Ratzeburgia*; **C**, *Thyrisa*, **1**, front view, **2**, back view; **D**, *Hackelochloa*, **1**, front view **2**, back view; **E**, *Ophiuros*; **F**, *Hyparrhenia*. **i** = internode; **p** = pedicel.

The *Ischaeminae* and *Rottboelliinae* comprise the second main line of development within the *Andropogoneae*. Here it is found that some of the potentialities of the rhachis-segment disseminule are realized, for pedicel, internode and lower glume of the sessile spikelet combine to form a protective capsule. At first the pedicel and internode merely reinforce the edges of the spikelet (*Ratzeburgia* Kunth is a particularly elegant example—Fig. 9/B); at a later stage the greatly thickened pedicel and internode form two sides of a 3-cornered box closed by the lower glume (Fig. 9/C); alternatively the box may be formed by the lower glume opposed by the flattened internode and pedicel side by side (*Hackelochloa* Kuntze is an extreme example—Fig. 9/D); ultimately the fertile spikelet is sunk in a cylindrical capsule formed by the fused pedicel and internode (Fig. 9/E). The lower glume is often armoured with ridged patterns; hairs and awns are rarely present; and the pedicelled spikelet, which cannot conveniently be enclosed in the capsule, is commonly suppressed. It is apparent that specialization has been almost wholly in favour of protection.

The third major line, the *Andropogoninae*, is rather variable. Some of the least modified genera (such as *Bothriochloa* Kuntze) tend to have small plumose wind-dispersed spikelets like those of the *Saccharinae*. In two sections of *Andropogon* there is a reincarnation of protective pedicels and internodes, but with interesting differences from the *Rottboelliiniae*. In the first place the lower glume does not make a conspicuous contribution to the structure, but instead tends to retreat with the rest of the spikelet into the cleft between pedicel and internode (is the grooved glume of *Diheteropogon* and allied genera merely a useless relict of this tendency?). Secondly the pedicelled spikelet, instead of degenerating, commonly becomes enlarged to form an imbricate covering for the back of the raceme. By and large, however, the pedicels and internodes of the *Andropogoninae* are slender, so that the disseminule consists essentially of a single spikelet protected by its indurated glumes (Fig. 9/F, p. 19).

In the *Andropogoninae* the sessile spikelet tends to become a specialized female organ, chiefly by loss of function in the lower floret. For example in *Hyparrhenia* the latter is represented only by a lemma, which apparently serves as a substitute for the absent palea of the upper floret. Perhaps by way of compensation, the lightly protected pedicelled spikelets are nearly always preserved as functioning male organs. In view of the fact that spikelet and disseminule are almost equivalent, it is not very clear why sexual dimorphism should be so strongly expressed, even in supposedly primitive genera such as *Sorghum* Moench. It may be that it was inherited from an ancestor (*Ischaeminae?*) with a box-type disseminule in which the character was already well established. Or it may be that the callus forms an essential part of the disseminule, and that this structure has evolved primarily by modification of the rhachis segment. A callus at the base of the pedicelled spikelet presumably evolved independently, and opened the way to a reversal of the sexual roles of sessile and pedicelled spikelets such as occurs in *Trachypogon* Nees and *Germainia* Bal. & Poitr.

Dispersal

To interpret the function of the disseminule it is customary to seek an explanation of its structure in terms of dispersal mechanisms. Such mechanisms may be highly developed in the *Andropogoninae*. Consider the downy windborne spikelets of some species of *Andropogon*, the pungent glume tips of *Cymbopogon refractus* (R.Br.) A. Camus, or the needle-like callus of *Heteropogon contortus* (L.) P. Beauv. ex Roem. & Schult. In this species the twisting awns unsheath the callus and poise it outwards, so that the spikelets form a malignant tangled knot perched on the top of the culm, ready, at the lightest touch, to thread themselves into one's clothing.

Yet in most genera the dispersal mechanism seems singularly ineffective. Thus in *Hyparrhenia* the lighter and fluffier spikelets have some capacity for wind transport, and those with stiff hairs or a pungent callus may catch in one's clothing, but in neither case do they seem particularly efficient. It has already been suggested that evolution in the *Rottboelliinae* has been guided by an overriding need for protection, and this appears also to be the case in *Hyparrhenia*. Indeed, a dominant species of continent-wide climax savanna has little use for long-distance dispersal to exploit new habitats, but it has a great need for protecting its seed both in the raceme and after shedding. It is difficult to assess the relative importance of adverse conditions on the raceme, but the young flower is certainly very vulnerable, and perhaps also the ripening grain which comes to maturity in the early part of the dry season (in northern Nigeria at this time the relative humidity may change from 60% to 6% overnight as the inter-tropical front migrates southward). The shed grain is exposed to annual grass fires, and may also be stimulated into premature germination by unseasonal rainstorms. The raceme-segment of the *Rottboelliinae* provides dual purpose protection both in the raceme and after shedding. In *Hyparrhenia* (and many other *Andropogoninae*) there is a division of labour, the immature flower being protected by the spatheole, while the awn provides the means for protecting the shed grain.

The awn

A prominent feature of the *Hyparrhenia* spikelet is its awn. This contains no chlorenchyma (Tran, 1965), and cannot therefore make a significant contribution to photosynthesis, as does the awn of *Triticum* Linn. (Grandbacher, 1963). The twisted column of the awn reacts to hygroscopic changes by coiling and uncoiling, a movement which may be of some value in freeing the grain from the leafy tangle of the inflorescence. Of greater importance is the ability of the flailing limb, in conjunction with the retrorsely bearded callus, to propel the grain along the ground. The distance involved is too small to constitute efficient dispersal (10–30 cm. after wetting and drying once daily for a week) but, most significantly, it always culminates in the burial of the spikelet. The depth of burial is slight, but then fire temperatures, which can be lethal a few millimetres above the surface, are scarcely felt just below it (Guilloteau, 1957). The spikelet of *Hyparrhenia* can thus be seen as a most successful response to the need to secure protection for the seed from the annual savanna fires. Burial may also help to delay germination until the soil is thoroughly soaked by rain, thus avoiding a false start following an out-of-season storm.

The significance of the various callus shapes found in *Hyparrhenia* is not clear; crude experiments suggest that they are equally efficient at burying the seed. The awn hairs may have some relation to the retention of dew, improving the efficiency of the hygroscopic mechanism in the early dry season when the grain is shed.

The inflorescence

If the success of the grasses were to be summarized in one word, that word would be leaf-sheath. Owing to the mechanical properties of the stem and sheath combination, the culm meristem can be located in a protected site at the base of the internode. Likewise the apical meristem can be remarkably well protected in a nest of concentric sheaths, through whose lumen the inflorescence is extruded at maturity. However this structure, together with the need to expose the inflorescence high in the air for wind pollination, imposes limitations, and the typical grass inflorescence is usually a single terminal panicle which may often be very large. The alternative, a number of smaller axillary inflorescences, has seldom been exploited (except, of course, in the form of tillers), but it is the evolutionary course taken by the *Andropogoneae*.

The apparently primitive genera of the *Andropogoneae* possess the customary terminal panicle. It may be supposed that the development of a bulky inflorescence within the confines of a concentric sheath system poses problems of protection during the later stages of maturation. At all events a progressive reduction of the inflorescence to 1 or 2 racemes can be observed, a process which renders both feasible and desirable a multiplication of the flowering system by means of axillary branching. This stage is well shown by many genera of the *Rottboelliinae*. The ultimate development, in *Hyparrhenia* and its allies, is a

concentration of axillary branching in the upper part of the culm and a reduction of the leafy component, producing a structure analogous to a true panicle
(Fig. 10). The wheel would appear to have turned full circle, but with the
important difference that when congestion occurs within the sheath there
is extruded, not an immature panicle, but a closed spathe within which further
expansion can take place. A sequence of such extrusions ensures that the
racemes are totally enclosed until their development is completed. Raceme
and spatheole thus become intimately associated as a single unit, and variation
within *Hyparrhenia* is essentially an exploration of the possibilities opened up by
this development, which bears a fascinating similarity to the process by
which the spikelet scales must have evolved from leaf-sheaths in the ancestral
grasses.

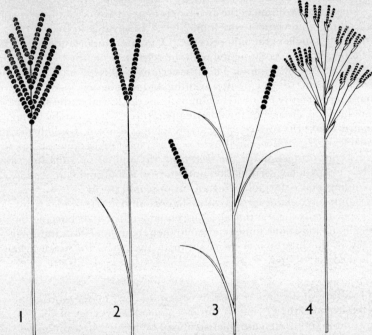

FIG. 10. Inflorescence types in *Andropogoneae*. **1,** large panicle (*Saccharum*); **2,** reduced panicle
(*Andropogon*); **3,** axillary branching (*Coelorhachis*); **4,** spathate panicle (*Hyparrhenia*).

The spatheole and raceme

In its simplest form the spatheole retains many characteristics of a leaf-
sheath, the many-awned racemes being terminally extruded at the tip of a long
peduncle (*e.g. H. hirta* and *H. rufa*). However the structure of the spatheole–
raceme unit permits the development of an alternative method of exsertion,
whereby exposure of the racemes is delayed until the last possible moment. It is
associated with abbreviation of the raceme and inflation of the spatheole, so
that the spikelets are wholly enclosed until they have completed their development, whereupon the racemes are thrust out laterally by deflexion of the
specialized raceme-bases (*e.g. H. cymbaria*) or, in one species (*H. madaropoda*),
by flexure of the peduncle.

Earlier, it was observed that the *Andropogoninae* show a progressive reduction
in the number of racemes, these being solitary in the more advanced genera.

This would appear to be the logical solution in the cramped conditions of a spathate panicle, where multiple racemes must suffer some degree of mutual interference. This tendency can indeed be observed in *Hyparrhenia*, where *H. mobukensis* displays an imperfect conversion to solitary racemes, and the long raceme-base of *H. filipendula* can be regarded as an attempt to secure a similar effect by placing the racemes end to end. Further development of this trend in *Hyparrhenia* is countered by the evolution of deflexing raceme-bases which serve to separate the racemes. It is of interest to note two other manifestations of the trend elsewhere in the tribe; the abortion of the upper raceme in *Hyperthelia colobantha*, and the interlocking of the two racemes back to back in species of *Ischaemum* to produce, in effect, a single composite spike.

There is always a homogamous pair at the base of the lower raceme; no doubt the reduction of this pair of spikelets is incidental to the modification of other parts of the lowest raceme-segment into raceme-bases. In passing it may be noted that the potential ability of raceme-segments to produce a second (branch) internode, may be invoked to explain the triads(2 sessile and 1 pedicelled spikelet at each node) that make up the racemes of *Mnesithea* Kunth, *Lasiurus* Boiss. and *Polytrias* Hack., by assuming that the 'branch' initial has here developed into a sessile spikelet. Two other patterns of homogamy may be observed in *Hyparrhenia*. The *Filipendula*-type (2 pairs at the base of the upper raceme) may be regarded as a further measure to eliminate mutual interference between the racemes. The *Diplandra*-type (2 pairs at the base of each raceme) are of an involucral nature, and seem designed to afford protection to the fertile spikelet in their midst. The eruption of bristles from the raceme-base of several species, but significantly not from those bearing involucral homogamous pairs, presumably has a similar function. It is fascinating to observe the parallel evolution of reduced racemes subtended by involucres and enclosed by inflated spatheoles in such diverse genera as *Apluda* L. (in which the involucre is derived from pedicels), *Hyparrhenia diplandra*, *Themeda* Forsk. and *Iseilema* Anderss.

It was suggested (Clayton, 1966e) that the large raceme-base appendages of *Hyperthelia* might help to protect the immature fertile spikelet. This explanation cannot be applied to *Hyparrhenia* sect. *Hyparrhenia* for, at all stages, the appendages are much shorter than the developing spikelet. It is possible that in this section the appendage, which is situated on the inner side of the raceme, helps to transfer the thrust of deflexion to the long callus of the sessile spikelet, and thus relieve stress on the fragile abscission line. It may be significant that the appendages are poorly developed in *H. madaropoda*, the only member of the section whose racemes remain permanently parallel. However the dubious validity of such reasoning is exposed by *H. andongensis*, which remains firmly anomalous.

GEOGRAPHICAL DISTRIBUTION

Any discussion of geographical distribution is confronted by a number of difficulties:

1. The boundaries of the most suitable geographical units (chorological domains) are rather loosely defined.
2. The distribution of the individual species is often imperfectly known.
3. The crude areal extent of a species may assume a different significance when compared with its ecological preferences, but these are seldom adequately recorded.

4. The absolute range of most species is rather wide. In order to make meaningful comparisons it is necessary to distinguish arbitrarily between the
main core of the distribution, and sporadic outlying records of much less
importance.

It is thus inevitable that a chorological discussion should deal with broad
generalizations; nevertheless it is possible to make useful deductions.

African distribution of the genus

All the species of *Hyparrhenia* occur in Africa and it is sufficient, in the first
instance, to concentrate upon that continent. For the choice of geographical
units White's (1965) map of chorological regions seems to be the most promising (Map 1), save that for the present purpose his Sudanian Domain has
been divided into eastern and western halves along a line through Lake Chad.

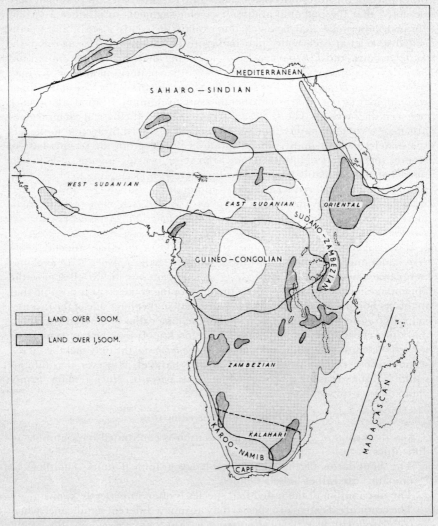

MAP 1. Chorological Regions (upright lettering) and Domains (sloping lettering) of Africa
with Afro-alpine and Afro-montane Regions omitted. Modified from White (1965).

With the exception of *H. hirta* (Mediterranean and Cape Regions), the genus is virtually confined to the Sudano-Zambezian Region. This is to be expected as this Region corresponds to a major ecological formation—the tropical savanna, whose extent is primarily controlled by climatic factors. The distribution of the species among the Domains of the Region is shown in Table 3, in which the minutiae of locality lists have been reduced to a manageable

TABLE 3. Distribution of the species of *Hyparrhenia* among the Domains of the Sudano-Zambezian Region. × = 'core' distribution. ○ = endemic.

Species	West Sudanian	East Sudanian	Oriental	Zambezian
H. glabriuscula	×			×
H. mobukensis			×	
H. nyassae		×	×	×
H. smithiana	○			
H. exarmata		×		
H. rufa	×	×	×	×
H. dichroa				×
H. poecilotricha			×	×
H. gazensis				×
H. finitima				×
H. wombaliensis		○		
H. hirta			×	×
H. quarrei			×	×
H. griffithii			×	
H. anamesa			×	×
H. violascens	×			
H. bagirmica	×			
H. barteri	×	×		×
H. familiaris		×		×
H. figariana		×		×
H. filipendula	×	×	×	×
H. andongensis				○
H. anthistirioides			×	
H. cymbaria		×	×	×
H. variabilis			×	×
H. pilgerana			×	
H. formosa			×	
H. schimperi			×	×
H. cyanescens	×			
H. papillipes			×	
H. dregeana			×	×
H. tamba			×	×
H. umbrosa		×	×	×
H. rudis			×	×
H. collina			×	×
H. madaropoda		○		
H. coleotricha			×	
H. confinis		○		
H. niariensis		×		×
H. welwitschii	×	×		×
H. bracteata		×		×
H. newtonii		×		×
H. multiplex			○	
H. anemopaegma				○
H. arrhenobasis			○	
H. tuberculata			○	
H. dybowskii		○		
H. involucrata	×			
H. mutica	×	×	×	×
H. subplumosa	×			
H. diplandra	×	×	×	×
H. gossweileri				×

form by considering only the 'core' distribution. It is a simplification which involves considerable subjective judgement, but which I believe represents a fair precis of the true situation. The records may be summarized as follows:

Regional: 4 species. It is remarkable that *H. rufa* is the only species common throughout the Sudano-Zambezian region. The other species included here are not very common in W. Sudan.

E. Sudan–Oriental–Zambezian: 3 species. These otherwise Regional species fail to extend into W. Sudan.

W. Sudanian: 6 species (including 1 endemic). The Domain is noticeably isolated from the rest of the continent. *H. rufa* and *H. welwitschii* are the only common W. Sudanian species to achieve any considerable distribution elsewhere. Conversely the otherwise Regional species are poorly represented; moreover two of them (*H. diplandra* and *H. nyassae*) are largely replaced west of the Cameroun mountain range by vicarious counterparts. It is probable that the Cameroun mountains and the Pleistocene enlargement of Lake Chad have partially isolated this Domain from the rest of the Region (Moreau, 1963); no doubt the flora was further improverished by the southward advance of the Sahara which occurred at some time in the late Pleistocene (Clayton, 1957 & 1966b).

E. Sudan–Zambezian (with W. Sudan extension): 8 species. Species which occur to the north and south of the Guinean Region, but are absent from the Oriental Domain. *H. welwitschii* and *H. barteri* extend into the W. Sudanian Domain; *H. glabriuscula* is centred there.

E. Sudanian: 5 species (including 4 endemics). Note the high degree of endemism, which might be still higher if the three species of *Arrhenopogonia* were included here rather than in the Oriental Domain. On the other hand the inclusion of *H. wombaliensis* is questionable.

Oriental: 10 species (including 3 endemics).

Zambezian: 6 species (including 2 endemics).

Oriental–Zambezian: 10 species. Note the disjunct distribution of *H. dregeana* and *H. tamba*, which are rare in the central sector of these domains; *H. hirta* shows the same disjunction in a more extreme form.

Distribution of the sections

Table 3 may be summarized in a different form to show the geographical distribution of each section of the genus:

I. *Strongylopodia*, 2 species. *H. glabriuscula* in W. Sudanian, with suggestions also of a Zambezian distribution. *H. mobukensis*, although nominally Oriental, is in fact confined to the mountain massifs.

II. *Polydistachyophorum*

    (a) *Rufae*, 8 species (including 1 endemic). May be loosely regarded as Regional, but with a strong Zambezian (3 species) bias.

    (b) *Hirtae*, 5 species (including 1 endemic). *H. hirta* has a disjunct distribution in the north and south of the continent; *H. anamesa* is Oriental–Zambezian; 2 species are Sudanian; and *H. wombaliensis* is isolated in the lower Congo basin.

    (c) *Filipendulae*, 6 species. One widespread species (*H. filipendula*), and 5 related species in the Sudanian Domain.

    (d) *H. andongensis*. Angola.

III. *Pogonopodia*, 12 species. Centred upon the Oriental Domain, but well represented in the Zambezian; only 1 species elsewhere (*H. cyanescens*).

IV. *Hyparrhenia*, 7 species (including 2 endemics). Apart from *H. coleotricha*, the section has an E. Sudan–Zambezian distribution, being ranged around the Congo basin with endemics in the E. Sudanian Domain.

V. *Arrhenopogonia*, 4 species (including 4 endemics). Ethiopia, with one species in Zambia.

VI. *Apogonia*, 6 species (including 1 endemic). Two regional species neither of them well distributed in the Oriental Domain; 3 Sudanian; and 1 Zambezian.

The conclusions to be drawn from the foregoing are that two broad patterns of distribution may be discerned in Africa. The one on the Western side of the continent (Sudan–Zambezian type), with a strong tendency to endemism in the E. Sudanian Domain. The other on the eastern side in the Oriental and Zambezian Domains. Further discussion of the significance of these distributions will be included in the section on phylogeny below.

TABLE 4. Extra-African distribution of species of *Hyparrhenia* and some allied genera

	Madagascar	Europe & Middle East	India & Ceylon	Indo-China	Indonesia	Australia	America
H. nyassae	×			×			
H. rufa	×			×		×	×
H. hirta	×	×	×			×	×
H. quarrei						×	
H. griffithii	×		×				
H. familiaris				×			
H. filipendula	×		×		×	×	
H. cymbaria	×						
H. variabilis	×				×		
H. schimperi	×						
H. papillipes	×						
H. rudis	×						
H. welwitschii	×						
H. bracteata							×
H. newtonii	×			×	×		
H. diplandra				×	×		
Elymandra							×
Exotheca				×			
Hyperthelia	×						×
Parahyparrhenia				×			

Extra-African distribution

The species occurring outside continental Africa are listed in Table 4. The Madagascan species presumably represent an impoverished sample of the mainland flora. The distribution of *H. hirta* in the Mediterranean, Middle East and India can be seen as an obvious consequence of an African species becoming adapted to a different climate. The presence of *H. hirta* and *H. quarrei* in Australia is probably due to introduction. This leaves two distributions in the Old World to be considered:

(a) *H. filipendula*. Its distribution follows a classical pattern—Madagascar, Ceylon, Indonesia and Australia. Thus, while one may be inclined to dismiss its distribution as due to introduction, second thoughts suggest that this may not have been so.

(b) Indo-China. No less than 5 species (6 if *H. griffithii* from Assam is included) occur in Indo-China, as also do the neighbouring genera *Exotheca* and *Parahyparrhenia*. The conclusion that Indo-China forms a natural extension of the African distribution is inescapable, and Schnell (1962) has discussed the reasons for believing that at one time species were free to migrate between here and Africa. Moreau (1952) observes that there is geological evidence that savanna species could pass between Africa, Arabia and India during the Miocene and early Pliocene, but he admits that an earlier connection would seem necessary to account for some animal and plant distributions. It should be remembered, however, that there has been an extensive commerce between eastern Africa and the Far East for nearly 1000 years (Freeman-Grenville, 1962), and the seeds of grasses and other weeds may have travelled the same route. The range of variation of the species in Asia is commonly much less than that of the African population.

Three species, all common in Africa, have entered the New World. *H. hirta* is so common in Spain and Portugal that it would be surprising if it had not been introduced to similar climates in the overseas possessions. The other two species, and particularly *H. rufa* which is commonly used as a general purpose straw, are equally likely to have been introduced at the time of the Slave Trade. One curious feature however is the distribution of the rare ally *Elymandra lithophila* in the Katangan region and in Brazil. Can this be taken as evidence that a few African species had become established in America before the institution of regular commerce across the Atlantic? However, Schnell (1961) observes that the floristic affinities between Africa and America are mainly at the level of family and genus, and he enjoins caution in the interpretation of specific identities. *Hyparrhenia* at least does not seem to have been differentiated from proto-*Andropogon* until after the separation of America from Africa.

Allied genera

Earlier in this review (p. 2 *et seq.*) I suggested a sequence of morphological progression within the *Andropogoneae*. It is noticeable that the most primitive sub-tribe, the *Saccharinae*, has a predominance of genera and species in the Asiatic tropics, and it may be concluded that the tribe is of Asiatic origin. In the sub-tribe *Andropogoninae* the primitive group *Bothriochloastrae* is likewise found to be predominantly Asiatic, but a productive burst in Africa has produced the greatest number of genera in the *Andropogoninae*. There is a complication in that the most advanced genera (*Themeda* and *Iseilema*) are once again Asian or Australian. Support can thus be found for the diphyletic theory in that the genera with entire lemmas are Asian, and those where they are bilobed are African.

Another significant point is that up to *Andropogon* and *Schizachyrium* there are a number of genera with a Gondwanaland distribution and endemic species in S. America, but beyond this point there are very few (and these from the *Heteropogonastrae*, the most primitive of the divergent lines). America probably became separated from Africa at the beginning of the Cretaceous. There is therefore evidence that the relative ages of genera in our morphological sequence are in accord with their evolution in a similar order.

The genera regarded as parental to *Hyparrhenia*, namely *Andropogon* and

Cymbopogon, have a pan-tropical or palaeotropical distribution. The genera allied to it, which are very rare outside Africa, have a marked bias towards a Sudanian distribution, with centres of endemism in Fouta Djallon and also around the frontier between the Sudan and Central African Republics. The phylogenetic implications will be explored further in the next section.

PHYLOGENY

Phylogeny is rightly treated with reserve by taxonomists, owing to its very high speculative content, and the tendency to circular reasoning latent in many of its arguments. This is certainly true of the *Andropogoneae*, where the sub-tribe *Saccharinae* provides a happy hunting ground for polyphyletic conjecture. Nevertheless the plants we have to classify are the product of phylogenetic processes, and it would be unrealistic to give no thought to the taxonomic implications. Fortunately, *Andropogonastrae*, *Hyparrheniastrae* and *Anadelphiastrae* together seem to form a monophyletic group, and, by thus restricting attention, the more extravagant hypotheses that a wider enquiry might evoke can be avoided. Let us therefore review such phylogenetic evidence as can be adduced from the previous sections.

Adaptation

The logical sequence of adaptations to environmental pressure proposed earlier suggests a first approximation to phylogeny (Fig. 11). But it cannot be assumed that advantageous character combinations arose in the most orderly sequence, for there is evidence that the primitive gene pool of *Hyparrhenia* already contained much potential variation.

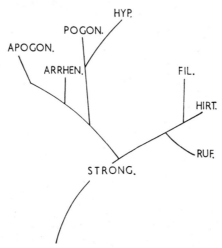

Fig. 11. Phylogeny of *Hyparrhenia*: evolutionary adaptation. Apogon. = *Apogonia*, Arrhen. = *Arrhenopogonia*, Fil. = *Filipendula*, Hirt. = *Hirta*; Hyp. = *Hyparrhenia*, Pogon. = *Pogonopodia*, Ruf. = *Rufa*, Strong. = *Strongylopodia*.

Morphological relationships

The starting point here must be the diagram of morphological relationships (Fig. 3, p. 5) which, by the criteria of increasing differentiation and elaboration of organs (discussed on p. 2 *et seq.*) proceeds phylogenetically from

Andropogon towards *Anadelphia*. The difficulty is that it displays a ring structure, an unlikely, though not impossible, phylogenetic pattern. When the extent to which each genus retains echoes of its predecessor and foreshadows its successor is considered, the weakest link in the chain is seen to be between *Hyparrhenia* and *Hyperthelia*. The supposed affinity between the two rests mainly upon the raceme-base appendages, although they are of a different order of magnitude in the two genera, and possibly also of different function; the analogy between the glumes of *Hyparrhenia newtonii* and atypical variants of *Hyperthelia dissoluta* rests on still less sure ground. Figure 3 might therefore be transformed into the phylogenetic diagram shown in Figure 12.

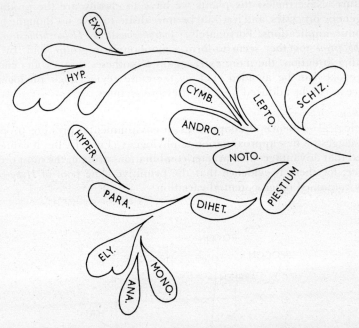

Fig. 12. Phylogeny of genera allied to *Hyparrhenia*. Ana. = *Anadelphia*, Andro. = *Andropogon* Sect. *Andropogon*, Cymb. = *Cymbopogon*, Dihet. = *Diheteropogon*, Ely. = *Elymandra*, Exo. = *Exotheca*, Hyp. = *Hyparrhenia*, Hyper. = *Hyperthelia*, Lepto. = *Andropogon* Sect. *Leptopogon*, Mono. = *Monocymbium*, Noto. = *Andropogon* Sect. *Notosolen*, Para. = *Parahyparrhenia*, Piestium = *Andropogon* Sect. *Piestium*, Schiz. = *Schizachyrium*.

Hyparrhenia itself appears to have arisen from the common stock of *Andropogon* and *Cymbopogon* before the characteristics (i.e. the shape of lower glume and raceme-bases) of the present genera had become firmly established. Sect. *Strongylopodia* probably lies close to the ancestral line for both its species have primitive characters (the *Cymbopogon*-like features of *H. glabriuscula*, and the glabrous awn of *H. mobukensis*).

Two further points emerge from a consideration of morphological affinity. Firstly, sects. *Polydistachyophorum* and *Pogonopodia* are so closely related morphologically that they can scarcely be unrelated phylogenetically. Secondly, sects. *Pogonopodia* and *Apogonia* have very little in common despite their flattened raceme-bases; indeed the latter bears a closer resemblance to *Exotheca*, which must presumably be regarded as older than any of the sections of *Hyparrhenia*. It seems therefore that sect. *Apogonia* should be regarded as phylogenetically primitive, rather than as an advanced development from sect. *Pogonopodia*, and figure 13 shows the phylogenetic diagram modified accordingly.

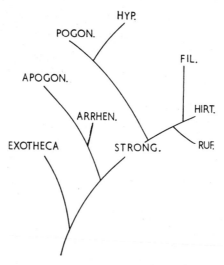

Fig. 13. Phylogeny of *Hyparrhenia*: modified according to morphological affinities. Apogon. = *Apogonia*, Arrhen. = *Arrhenopogonia*, Fil. = *Filipendula*, Hirt. = *Hirta*, Hyp. = *Hyparrhenia*, Pogon. = *Pogonopodia*, Ruf. = *Rufa*, Strong. = *Strongylopodia*.

Geographical distribution

In the absence of fossils the only chronological evidence available is that provided by the present day distribution of species, although such evidence is indirect and difficult to interpret. The high degree of endemism in the Sudanian Domain, both within *Hyparrhenia* and among its allies, suggests that the genus originated in this area.

It was suggested in the section on geographic distribution that two main patterns could be discerned. It can be contended that the first of these, the Sudan–Zambezian type, which is shared by several closely allied genera and which includes the E. Sudanian centre of endemism, is of a relict nature and its constituents phylogenetically older. The second pattern, the Oriental–Zambezian type, can then be seen as a later bout of vigorous speciation which has pressed the older species into the fringes of the Sudano-Zambezian Region. It is of interest to note that *H. hirta* and *H. dregeana*, morphologically the most primitive members of the second group, display a disjunct distribution, suggesting that they have been driven outwards by further speciation at the centre.

Such a hypothesis confirms the antiquity of sects. *Apogonia* and *Strongylopodia* (although of Oriental distribution, *H. mobukensis* has retreated to the mountain massifs), and suggests that the *Filipendulae* may also be a fairly primitive group. This is a satisfactory finding in view of the apparently paradoxical combination of involucral spikelets with end to end racemes in *Exotheca*, suggesting that both tendencies were present in the ancestral gene pool (but *Exotheca* itself has an Oriental–Zambezian distribution, though mostly in the highlands). Another important inference is that sect. *Hyparrhenia* also diverged

at a comparatively early date. Its similarity to *Hyperthelia* now becomes understandable, and even *Hyparrhenia andongensis* seems less of an anomaly. The final
modification to the phylogenetic tree is shown in figure 14.

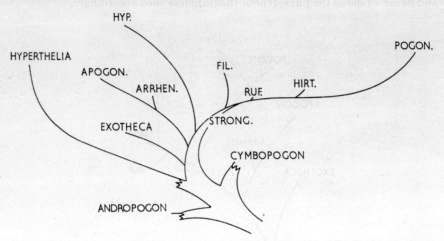

FIG. 14. Phylogeny of *Hyparrhenia*: modified according to geographical distribution. Apogon. =
Apogonia, Arrheno. = *Arrhenopogonia*, Fil. = *Filipendula*, Hirt. = *Hirta*, Hyp. = *Hyparrhenia*,
Pogon. = *Pogonopodia*, Ruf. = *Rufa*, Strong. = *Strongylopodia*.

Climatic change

In the foregoing discussion it has been tacitly assumed that evolutionary
changes in the genus have been generated by genetic forces operating within
it. This is only partially true, for the effect of external changes must also be considered. Those due to gross alterations in the configuration of the land surface
may be discounted, for the pre-Cambrian bulk of the continent is undoubtedly
a very ancient land mass which has suffered no appreciable marine trans-
gression since Cretaceous times.

Climatic change is a different matter, for there is abundant evidence that
during Tertiary times the African flora and fauna underwent many vicissitudes
as a result of climatic fluctuations. During most of this period the evidence is
fragmentary, so that it is not yet possible to perceive a coherent pattern, nor to
predict the repercussions that may persist in the present day taxonomy of the
genus. However, there is more extensive evidence for climatic conditions
during the Pleistocene, and it is certainly significant.

Moreau (1963) concludes that during the last glaciation (50,000–18,000
years ago) temperatures in Africa were some 5°C lower. This, together with
the concomitant increase in effective precipitation due to reduced evaporation,
would bring the lower limit of the upland evergreen forest from its present
1500 m. down to 500 m. (Map 1, p. 24). The surface contours are such that
upland forest would then cover the greater part of the continent, savanna
being largely confined to the present Sudanian Domain and the Congo basin,
and the lowland rain forest clinging precariously to limited stretches of the
west coast. It is suggested that it is the Pleistocene savannas which can be seen
reflected in the present Sudan–Zambezian distribution. A few species (notably
H. hirta) probably became adapted to the more arid climate which seems to
have persisted on the fringe of the savanna in the Ethiopian region, and which
must also have existed on its southern margin where the Kalahari desert

reached up into what is now the Congo forest. Ethiopia itself seems to have suffered little climatic disturbance, and may have been accessible at this time to some of the savanna species, such as the three species of sect. *Arrhenopogonia* now found there as endemics.

With a return to warmer conditions and the retreat of the Kalahari, the lowland rain forest expanded enormously, and the species of *Hyparrhenia* existing at that time were driven back to substantially their present distribution (*H. wombaliensis* being enveloped by the forest). At the same time the upland evergreen forest retreated, leaving vast stretches of land to be occupied by savanna. It seems that the invasion was accomplished, not so much by the pre-existing species, but by the eruption of a new genetic stock which eventually gave rise to the species which now display an Oriental–Zambezian distribution.

Conclusions

1. Sects. *Strongylopodia*, *Apogonia* and *Hyparrhenia* can be visualized as the broken remnants of older groups; *Polydistachyophorum* as a younger complex now beginning to disintegrate, the successful species becoming dominant and the less successful declining, but with speciation still active in places; and *Pogonopodia* as a vigorous new line in the full flush of speciation with, as yet, little sign of selective extinction.

2. In retrospect it can be seen that the classification of the genus rests upon a secure foundation. To the morphological characters of raceme-base and spatheole from which it is derived can be attributed a role of fundamental biological importance in adapting the plant to life in the savanna environment. The characters are subject to intense selection pressure exerted by the environment, and can be accepted as reasonably faithful indicators of the gene pool upon which it operates.

3. The constituent species of each section are so interrelated as to form a coherent whole, and it cannot be doubted that by and large the sections are natural entities.

4. The relationship between sections is more debatable. It has been customary to arrange them in a logical sequence of progressive modification. Such phylogenetic evidence as it has been possible to derive suggests that this view is mistaken, and that the flattened raceme-base, for example, has developed independently in three separate lines. It is a temptation to modify the taxonomy in accordance with phylogenetic hypotheses, but in the formal classification I have preferred to adhere to the straightforward logic of a natural morphological presentation, rather than to pursue the will-o'-the-wisp of phylogenetic conjecture.

ECONOMIC USES

Culwick (1950) reports that the leaves of *Hyparrhenia* are sometimes eaten by inhabitants of the Sudan. The grains of the related *Hyperthelia edulis* (C. E. Hubbard) W. D. Clayton are regularly collected for food (Hubbard, 1951). *Hyparrhenia filipendula* can furnish a paper-making pulp of moderate quality (Imperial Institute, 1935). The most important direct use of *Hyparrhenia* however is as a thatching material, for which purpose it is widely used throughout Africa.

Hyparrhenia rufa has acquired some reputation as a sown pasture grass in tropical America (Kemp *et al.*, 1961), but elsewhere it is usually regarded as inferior to a number of other grasses of proven worth. In general the tufted habit of *Hyparrhenia* does not form a satisfactory sward, and the plants soon grow away from the cattle to produce stemmy growth of little fodder value.

The economic interest of *Hyparrhenia* lies chiefly in its importance for extensive grazing on the open range. *Hyparrhenia* grasslands usually produce a low quality herbage, whose digestible dry matter and palatibility drop sharply at the beginning of the dry season. Surplus production cannot therefore be economically conserved, and cattle numbers are usually limited to the dry season carrying capacity. Despite these limitations there is no evidence yet that much of the natural range can be profitably replaced. However the greatest obstacle to range improvement is the considerable socio-political problem of securing effective control over grazing intensity.

NOMENCLATURAL HISTORY

Andropogon *L.*, Sp. Pl. 1045 (1753) & Gen. Pl. ed. 5: 468 (1754).

The first species of *Hyparrhenia* to be described were placed in *Andropogon*, a genus which originally embraced the greater part of the *Andropogoneae*. This enormous and unwieldy genus has gradually been dismembered, but it was as *Andropogon* that Hackel (1889) treated our genus in his great monograph. He did however recognize the taxonomic coherence of *Hyparrhenia* by regarding it as a distinct section. Stapf (1898) likewise employed *Andropogon* in a wide sense for his first major work on the group, as did the majority of authors up to the beginning of this century.

A number of species were described by A. Richard, who adopted the names originated by Hochstetter on printed labels distributed with exsiccatae from Schimper's Abyssinian collections. For the most part the labels bear no more than a name, date and locality, and the conventional citation 'Hochst. ex A. Rich. (1851)' is quite correct. However the original name was sometimes accompanied on the label by a brief printed Latin diagnosis, and must then be accepted as validly published on the label some seven years before the appearance of Richard's book.

A number of new names and combinations, ascribed to Stapf, were published by Chevalier. These names were copied from determination lists supplied to Chevalier by Stapf, who evidently regarded many of them merely as provisional names. Fortunately they have little influence on present nomenclature, and a critical evaluation of their status is unnecessary.

Vanderyst conducted investigations on Congo grasses during the first World War, publishing his findings soon after the Armistice. Being cut off from the libraries and herbaria of Europe, he was unable to compare his species with earlier work, and stated that his new names (which he marked with an asterisk) should therefore be regarded as provisional. Subsequent attempts by himself and others to validate the names or reduce them to synonymy have led to a good deal of confusion, now disentangled in the catalogue prepared by Compère (1963).

Anthistiria *Linn. f.*, Amoen. Acad. 10: 38 (1779).

Several species of *Hyparrhenia* have been placed in this genus, which is now reduced to a synonym of *Themeda* Forsk. The latter bears a remarkable resemblance to some species of *Hyparrhenia* (although the fertile lemma is entire and the racemes solitary), and it is understandable that the two genera should be confused at a time when the critical details of raceme structure were imperfectly understood.

Cymbopogon *Spreng.*, Pl. Pugill. 2: 14 (1815).

Sprengel transferred four species of *Andropogon* to his new genus, at the same time creating three superfluous names. The species (and basionyms) are: *C. elegans* (*A. cymbarius* L.), *C. glandulosus* (*A. prostratus* Willd.), *C. schoenanthus* (*A. schoenanthus* L.), and *C. humboldtii* (*A. bracteatus* Humb. & Bonpl. ex Willd.). Both the discussion in the protologue on the taxonomic position of *A. cymbarius*, and the etymology of the generic name, suggest that this species was foremost in Sprengel's mind and that it should have been taken as the type of the genus. On the other hand the generic description is equally applicable to all the species. There is therefore insufficient reason for upsetting the choice of *C. schoenanthus* as lectotype, implied by the combinations *Iseilema prostrata* (Willd.) Anderss. (1856), *Hyparrhenia cymbaria* (L.) Stapf (1918), and *H. bracteata* (Humb. & Bonpl. ex Willd.) Stapf (1918), and now entrenched by established usage.

Cymbopogon was generally reduced to a subgenus (Hackel, 1889) or section (Bentham & Hooker, 1883) of *Andropogon*, but was taken up at generic level by Rendle (1899) to include both *Cymbopogon* and *Hyparrhenia*. It was subsequently adopted by several authors describing new species prior to 1918.

Hyparrhenia *Fourn.*, Mex. Pl. 2: 51, 67 (1886).

To the first part of his 'Monographiae Andropogonearum' dealing with the *Anthistireae*, Andersson (1856) appended a list of excluded species beginning with the words '*Anthistiria pseudocymbaria* Steud. = est *Hyparrhenia* species', and followed by five other species to be transferred to *Hyparrhenia* (*Anthistiria quinqueplex* Hochst., *A. dissoluta* Nees, *A. multiplex* Hochst., *A. reflexa* H.B.K. and *A. foliosa* H.B.K.). It is obvious that he intended to return to the matter in a later paper, when the new genus would be described and the combinations made, but the work was never continued. A number of combinations ascribed to Andersson were effectively published by Schweinfurth (1867) in a list of Ethiopian plants. They were of course invalid, as the genus had yet to be described, but they are the reason for the citation 'Anderss. ex . . .'.

The genus, ascribed to Andersson, was eventually published by Fournier in 1886; Bentham and Hackel received sets of printer's signatures of his work in 1881, but these were given as personal favours, and cannot be accepted as equivalent to publication. The diagnosis of the genus was effected in a key, the relevant dichtotomy being:

'Spiculis geminis vel terminalibus tantum ternis .   . [other genera]
Spiculis numerosis polygamis fasciculatis   .   .   . Hyparrhenia.'

He included two species, *H. ruprechtii* and *H. foliosa*, and his inaccurate diagnosis would apply equally well to both, for in neither case is the paired nature of the spikelets obvious. With the transfer of *H. ruprechtii* to *Hyperthelia* (Clayton, 1966e), *Hyparrhenia foliosa* becomes the type species of the genus. Hitchcock's (1935) choice of *H. pseudocymbaria* as type, this being the first species listed by Andersson, is quite without foundation. Mansfeld's (1959) choice of *H. ruprechtii* is illogical, for he ascribed the genus to Stapf who founded his concept upon the section *Eu-Hyparrhenia* (= sect. *Polydistachyophorum*).

Hyparrhenia was first adopted extensively by Stapf (1918), and his work has remained substantially unaltered to the present day.

The most recent revision is that due to Roberty (1960) in his monograph on the *Andropogoneae*. He has wedded formal taxonomy to the concept of life-form index, from which a system of classification is developed by arithmetical manipulation. Although plausible, his premises are arbitrary and unproven, and they repudiate the genetic concept of species which guides most contemporary botanists. His classification is eccentrically novel and quite unacceptable, but nevertheless his new binomial combinations are apparently made in accordance with the International Code. However, he states (p. 25) that his infraspecific taxa are 'des groupements temporaires de formes incomplètement définies, non pas successivement subordonnées . . .'. This is a direct contradiction of Article 2, and it is apparent that Roberty is using these ranks in a sense to which the Code does not apply. I have accordingly ignored his nomenclatural innovations below the rank of species. He has also created two new sections (*Gryllopsis* and *Leptochaeta*) without Latin description or the designation of type species; these I have likewise ignored.

Other genera

A few combinations have been made with *Heteropogon*, *Trachypogon* and *Androscoepia* for no very good reason. A more extensive series of combinations were made in *Sorghum* by Kuntze (1891) who, at a time when rules of nomenclature were more open to subjective interpretation, concluded that *Sorgum* L. (1737) was available for priority.

SYSTEMATIC TREATMENT

The concept of species

In *Hyparrhenia* the concept of species is peculiarly difficult, for variation is apparently continuous throughout much of the genus. Both the available cytogenetic knowledge of the genus, and its successful occupation of an ecosystem constantly liable to disturbance, offer an explanation as to why this should be so. It should be appreciated that herbarium collections are likely to over-emphasize the degree of continuity; partly because the stable populations tend to be more distant from human occupation and therefore more trouble to collect, and also because their very uniformity renders them less interesting to the collector. To the taxonomist, who must catalogue variation in terms of discrete units, it is a thought which brings some comfort, but nevertheless he is faced with a formidable dilemma.

In such a situation there are three viewpoints from which the concept of species may be approached:

1. Intergradation is an indication that taxa are not distinct and should be united. When faced with indistinct boundaries such an argument has its attractions; but in this genus its impracticability soon becomes obvious as successive elements coalesce into a single species of unacceptable amplitude, unless capricious exceptions to the argument are admitted.

2. The nature of taxa and the degree of relationship between them should be expressed by a hierarchy of categories such as species, sub-species, variety and hybrid. Although an estimable solution in theory, it is at present impossible to provide consistent and workable definitions for the categories in

Hyparrhenia. The approach is soon found to degenerate into the speculative interpretation of morphological variation patterns, and it is pointless to construct an illusorily elaborate hierarchy from mere guesswork. It is, however, a development to be expected in the future as genecological knowledge of the genus expands.

3. Morphologically separable populations should, in general, be treated as species. The arbitrary allocation of specific rank is questionable, but it offers a practicable solution in a genus notorious for its lack of clear infra-generic disjunctions. It produces species of roughly comparable amplitude, which approximate closely to the largely intuitive classification proposed by earlier workers. This is the approach I have adopted.

The continuum concept, borrowed from ecology, is helpful when dealing with entities whose discontinuity is not readily ascertained. The specimens are visualized as clustered about a number of noda in a multidimensional network. The more concentrated noda are readily identifiable as species, and their density renders them susceptible to attempts to define their limits. This done, they provide the standards by analogy with which the more diffuse noda may be classified. Fundamentally, the taxonomy of *Hyparrhenia* hinges upon finding rational circumscriptions for a limited number of dominant species. Once these have been contained the classification of the remainder becomes a feasible proposition.

Subordinate categories

Despite the above remarks on taxonomic hierarchies, there is a case for recognizing an infra-specific category differentiated by minor, and often indistinct, morphological characters, usually associated with partial geographical or ecological separation. They have been termed varieties, although it is arguable whether subspecies would be more appropriate. They are of slight taxonomic importance, but have some interest as possible indicators of clinal variation, incipient segregation, or introgression.

Where necessary the species diagnoses have been supplemented by the description of 'minor variations'. They are sporadic deviations or specimens from the extreme range of normal variation, whose inclusion in the general diagnosis would widen it to a misleading degree.

Dimensions

A major difficulty in *Hyparrhenia* arises from the enormous variability of measured characters even within the same panicle. With specimens on herbarium sheets it is impracticable to try to reduce the variation by specifying precisely the location in the panicle of the part to be measured. The best that can be done is to select for measurement a spatheole (and subtended racemes) judged to represent the average condition within the panicle.

Variation within the species as a whole is even greater. Odd specimens with unusual dimensions can always be found, and to quote extreme ranges would introduce an element of anarchy into keys. The tails have therefore been eliminated when quoting range of variation. Where this may result in oversimplification, the true picture is displayed in scatter diagrams.

Spikelet measurements include the callus, but exclude fine mucros at the glume tip. Spatheole measurements exclude the filiform tip, if any, as this is so often broken off.

Bibliography

Most treatments of the genus in floras and plant lists stem directly from one of the three major studies (Hackel, 1889; Stapf, 1898 & 1918). Exhaustive bibliographic citations of works which contribute nothing new to knowledge of the genus have not therefore been attempted, especially as the identification of the specimens in such works is often most doubtful and impossible to verify.

The illustrations cited are a personal selection of high quality botanical plates. The citations are by no means exhaustive, and many other figures of less merit have been published.

Chromosome counts

Bibliographic references are given for counts taken from the literature, otherwise they are new. In both cases confirmation of voucher specimens is indicated by the sign !. It is obvious that the identification of species reported in the literature must often be treated with reserve.

Types

When types have been seen, the herbarium in which they have been located has been noted, using the abbreviations of the 'Index Herbariorum'. Otherwise the species have perforce been identified from descriptions, figures, or contemporary usage. Lectotypes have been selected wherever discordant elements are involved, and also where the syntypes are confusingly numerous.

Citation of specimens

It is impracticable to cite all the specimens examined, as the numbers were sometimes rather large (over 500 in the case of *H. rufa*). An arbitrary limit of about five for each country has therefore been set. Unless otherwise stated, the specimens cited are in the Kew Herbarium. (See p. 179 *et seq.*)

Sequence of species

I have followed convention by arranging the species in a sequence roughly corresponding to a supposed morphological progression. This has the disadvantage of placing the most difficult taxonomic groups at the beginning, and the reader may find it easier to tackle the sections in reverse order.

TAXONOMY

Hyparrhenia *Anderss. ex Fourn.*, Mex. Pl. 2: 51 (1886); Anderss. in Nov. Act. Soc. Sci. Upsal., ser. 3, 2: 254 (1856), sine descr.; Anderss. in Schweinf., Beitr. Fl. Aethiop.: 310 (1867), *sine descr.*; Stapf in Prain, Fl. Trop. Afr. 9: 291 (1918); Pilger in Engler, Nat. Pflanzenfam., ed. 2, 14e: 172 (1940). Type species: *H. foliosa* (H.B.K.) Fourn. (=*H. bracteata* (Humb. & Bonpl. ex Willd.) Stapf).

Andropogon L. subgen. *Cymbopogon* Nees, pro parte; Benth. & Hook. f, Gen. Pl. 3 (2): 1134 (1883); Stapf in Thistleton-Dyer, Fl. Cap. 7: 336 (1898).

Andropogon subgen. *Cymbopogon* sect. *Hyparrhenia* (Fourn.) Hack. in DC., Monogr. Phan. 6: 617 (1889).

Dybowskia Stapf in Prain, Fl. Trop. Afr. 9: 382 (1918). Type species: *D. seretii* (De Wild.) Stapf (=*Hyparrhenia dybowskii* (Franch.) Roberty).

Perennials or annuals, typically forming clumps of stout erect culms 1–3 m. high, rarely trailing. *Leaf-blades* linear, never aromatic; ligules scarious, truncate or rounded. *True inflorescence* represented by a pair of racemes, each supported by a short raceme-base and together borne upon a common peduncle, the latter filiform and subtended by a spatheole; spatheoles aggregated into an often elaborate spathate false panicle. *Racemes* comprising 1 or more pairs of spikelets, one of each pair sessile the other pedicelled (but the terminal sessile spikelet with 2 pedicelled attendants); raceme-bases often deflexed at maturity, the lower seldom over 1 mm. long, the upper often longer; pedicels and rhachis internodes filiform, ciliate both sides, shorter than the spikelets. *Homogamous pairs* 1–2 at the base of the lower raceme, 0–2 at the base of the upper; spikelets male or barren, muticous but otherwise similar to the pedicelled spikelets, tardily disarticulating after the other spikelets have fallen. *Sessile spikelets* 2-flowered, produced at the base into a blunt or pungent bearded callus obliquely set upon the internode below, disarticulating with contiguous internode and accompanying pedicel; lower glume as long as the spikelet, chartaceous, broadly convex across the back, the sides narrowly involute in the lower two-thirds, obscurely keeled above, narrowly truncate or bidentate at the tip; upper glume obtuse to acute. *Lower floret* reduced to a hyaline, 2-nerved lemma with ciliolate margins; palea absent. *Upper floret hermaphrodite*; lemma stipitiform, hyaline below, gradually hardening above with hyaline margins and bidentate tip, passing between the teeth into a stout geniculate awn with spirally twisted, pubescent or hirtellous column (rarely awnless); palea absent or minute; lodicules 2, cuneate, truncate or emarginate; stamens 3; stigmas 2, laterally exserted; caryopsis oblong, subterete, with basal hilum and embryo $\frac{1}{3}$–$\frac{1}{2}$ as long as the grain. *Pedicelled* spikelets acute, often aristulate from the lower glume, usually longer than the sessile; lower floret reduced to a hyaline lemma; upper floret male or barren with a hyaline lemma and no palea; pedicel tip often produced into a short triangular or subulate tooth appressed to the base of the upper glume.

The overall classification of the genus is displayed diagrammatically in Figure 15 (p. 40).

Keys to the sections and species of Hyparrhenia

Callus of sessile spikelet broadly rounded, semicircular; spikelets glabrous; raceme-bases unequal, the upper 1·5–3 mm. long
>> I. **Strongylopodia** (p. 47)

Callus of sessile spikelet acute to pungent, rarely obtuse but then the other characters not as above:
> Upper raceme-base terete or filiform, usually much longer than the lower, glabrous or sometimes softly hirtellous (stiffly pilose in *H. gazensis*), not produced into a scarious appendage (except *H. andongensis*); spatheoles generally narrow . . . II. **Polydistachyophorum** (p. 50)
> Upper raceme-base flattened, usually not much longer than the lower:
>> Homogamous spikelets at the base of the lower raceme only; raceme-bases bearded:
>>> Raceme-bases unappendaged, though sometimes with a scarious rim (if with a short appendage then racemes more than 4-awned per pair); upper raceme-base seldom over 1·5 mm. long
>>>> III. **Pogonopodia** (p. 102)

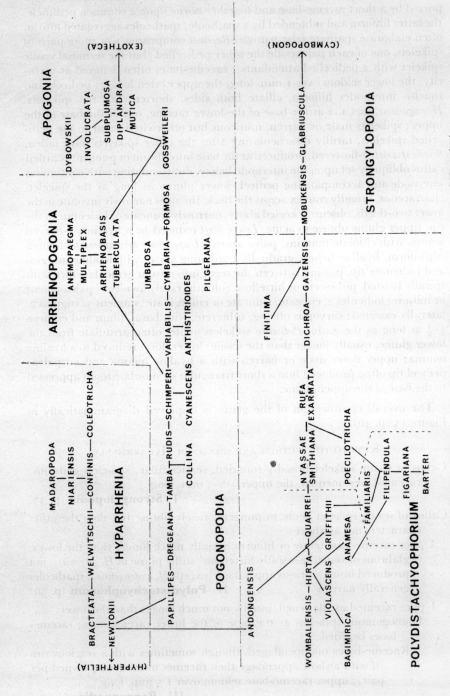

Fig. 15. Diagram showing morphological relationships among the species of *Hyparrhenia*.

Raceme-bases extended into a scarious appendage 0·5–4 mm. long;
racemes 2–4-awned per pair (more in *H. coleotricha*)
IV. Hyparrhenia (p. 132)

Homogamous spikelets at the base of both racemes:

Homogamous spikelets pectinate-ciliate on the margin; raceme-bases
usually bearded   .   .   .   .   V. Arrhenopogonia (p. 151)

Homogamous spikelets scabrid on the margin; raceme-bases glabrous
VI. Apogonia (p. 157)

I. SECT. STRONGYLOPODIA

Plant robust, caespitose; leaf-blades up to 30 cm. long; panicle copious; awn
puberulous .   .   .   .   1. glabriuscula (p. 47)

Plant slender, trailing; leaf-blades 4–8 cm. long; panicle scanty; awn glabrous
2. mobukensis (p. 48)

II. SECT. POLYDISTACHYOPHORUM

Spikelets wholly or partly rufously hairy:

Basal leaf-sheaths densely pubescent to tomentose below:

Sheath hairs white; callus of sessile spikelets 0·8–1·2 mm. long, acute;
beard at foot of raceme-bases usually white .   3. nyassae (p. 53)

Sheath hairs dark purplish red; callus of sessile spikelets 0·5–0·8 mm. long,
obtuse to narrowly truncate at the tip; beard at foot of raceme-bases
normally fulvous .   .   .   .   4. smithiana (p. 57)

Culms 40–100 cm. high; spikelet indumentum very dense, dark rufous
to chocolate coloured .   .   .   .   4a. var. smithiana (p. 57)

Culms 150–250 cm. high; spikelet indumentum rufous or fulvous
4b. var. major (p. 57)

Basal leaf-sheaths glabrous:

Fertile spikelets, or some of them, awnless, 12–19 per raceme-pair
5. exarmata (p. 60)

Fertile spikelets awned:

Sessile spikelets 3–5 mm. long; pedicelled spikelets awnless, rarely
mucronate:

Peduncles exceeding the spatheoles at maturity, at least on the periphery of the panicle; spatheoles linear, 3–5 mm. long;
raceme-bases glabrous .   .   .   .   6. rufa (p. 60)

Racemes (7–)9–14-awned per pair; upper raceme-base 2–3·5
mm. long, with or without a homogamous pair; callus obtuse,
0·2–0·8 mm. long .   .   .   .   .   6a. var. rufa (p. 62)

Racemes 6–7-awned per pair; upper raceme-base 3–5 mm. long,
with 1–2 homogamous pairs; callus slender, narrowly truncate, 1–1·2 mm. long .   .   6b. var. siamensis (p. 66)

Peduncles half to as long as the spatheoles, mostly 1–2·5 cm. long; spatheoles narrowly lanceolate, 2–3·5 cm. long, embracing the racemes; upper raceme-base 1·5–2·5 mm. long, usually hirtellous; racemes 6–9-awned per pair; callus cuneate, obtuse

7. **dichroa** (p. 68)

Sessile spikelets 5·5–7 mm. long; upper raceme-base 3·5–7 mm. long, with 1–2 homogamous pairs; pedicelled spikelet often with an awn-point up to 2 mm. long; racemes 4–7-awned per pair, terminally exserted; callus acute to pungent, 1–2 mm. long

8. **poecilotricha** (p. 69)

Spikelets white hairy or glabrous:

Upper raceme with or without 1 homogamous pair:

Plants perennial:

Spikelets glabrous to hispidulous; racemes not deflexed:

Raceme-bases produced into an oblong appendage 1 mm. long; lower glume of sessile spikelet longitudinally ridged

22. **andongensis** (p. 102)

Raceme-bases without appendages; lower glume of sessile spikelet not ridged:

Inflorescence copious; racemes 2–6-awned per pair:

Culms slender; callus cuneate, 0·8–1·5 mm. long; upper raceme-base (2–)2·5–3·5 mm. long, stiffly pilose

9. **gazensis** (p. 71)

Culms robust; callus linear, 1–2 mm. long; upper raceme-base 1·5–2·5 mm. long, hirtellous . . 10. **finitima** (p. 72)

Inflorescence of 1–2 raceme-pairs; racemes 9–12-awned per pair; raceme-base glabrous . . . 11. **wombaliensis** (p. 75)

Spikelets pubescent to villous:

Racemes not deflexed . . . . . . 12. **hirta** (p. 75)

Racemes, or some of them, deflexed:

Upper raceme-base 2–3·5 mm. long; rhachis internodes 2–3 mm. long; sessile spikelets 4·5–5·5 mm. long; awn 1·8–3·6 cm. long

13. **quarrei** (p. 82)

Upper raceme-base 3·5–8 mm. long; rhachis internodes 3·5–4·5 mm. long; sessile spikelets 6–7 mm. long; awn 4–6 cm. long

14. **griffithii** (p. 84)

Plants annual:

Racemes 6–9-awned per pair; column of awn bearing hairs up to 1·5 mm. long; upper raceme-base 2–3 mm. long:

Awn 2·5–4 cm. long, pubescent with hairs 0·1–0·5 mm. long; callus 0·8–1·2 mm. long, acute . . . . 16. **violascens** (p. 88)

Awn 5–8 cm. long, hirtellous with hairs 1–1·5 mm. long; callus 1·5–2·5 mm. long, pungent . . . . 17. **bagirmica** (p. 89)

Racemes 2-awned per pair; awn hirsute with hairs 3–5 mm. long; upper raceme-base 4–7 mm. long . . . 18. **barteri** (p. 90)

Upper raceme with 2 homogamous pairs:

Peduncle with spreading yellow hairs towards the top; racemes deflexed; awns 3–5 per raceme-pair; pedicelled spikelet with an awn 2–10 mm. long   .   .   .   .   19. **familiaris** (p. 92)

Peduncle with or without white hairs towards the top; racemes not deflexed; pedicelled spikelet with an awnlet up to 5mm. long:

Annual   .   .   .   .   20. **figariana** (p. 94)

Perennials:

Awn 3–5·5 cm. long, hirtellous with hairs 0·7–1·2 mm. long; callus 1·8–3 mm. long; raceme-base (4–)4·5–8(–10) mm. long; pedicelled spikelet with an awn 1–5 mm. long

21. **filipendula** (p. 95)

Spikelets glabrous; racemes 2-awned per pair

21a. **var. filipendula** (p. 95)

Spikelets white villous; racemes 2–4-awned per pair

21b. **var. pilosa** (p. 97)

Awn 2·5–4 cm. long, pubescent with hairs 0·1–0·6 mm. long; callus 1–1·8 mm. long; raceme-base 3·5–6 mm. long; pedicelled spikelet with or without an awn-point up to 2 mm. long; spikelets white villous; racemes 4–6(–7)-awned per pair

15. **anamesa** (p. 85)

III. Sect. Pogonopodia

Pedicelled spikelets glabrous (sometimes sparsely pilose in *H. schimperi*):

Plant annual; spatheoles 1·8–3·2 cm. long; awns 3·2–4·5 cm. long

23. **anthistirioides** (p. 106)

Plants perennial:

Callus square or broader than long, broadly obtuse; spatheoles 0·8–1·8 (–2·1) cm. long; awns 3–5(–6) per raceme-pair, up to 1·6(–2) cm. long; peduncle 3–8 mm. long; plant robust, rarely slender

24. **cymbaria** (p. 110)

Callus oblong to cuneate, obtuse or acute (sometimes square but then spatheole longer than above):

Awns 3–5 per raceme-pair; spatheoles 1·4–2·4 cm. long; awns 1·8–3·2 cm. long; peduncles 3–9 mm. long; callus cuneate; culms stout

25. **variabilis** (p. 113)

Awns 6 or more per raceme-pair in at least part of the panicle, rarely fewer but then plant slender and rambling:

Plant slender and rambling; spatheoles 2–3 cm. long; awns 0·7–1·7 cm. long; peduncles 9–30 mm. long, often sinuous, racemes commonly exserted; callus square to oblong, broadly rounded

26. **pilgerana** (p. 115)

Plant robust:

Awns 10–25 per raceme-pair, seldom over 2 cm. long; spatheoles 2·5–5 cm. long; peduncles 15–50 mm. long; racemes dense, the internodes 1·5–2 mm. long   .   31. **dregeana** (p. 124)

Awns 6–8 per raceme-pair, sometimes more but then the awns over 2·5 cm. long:

Awns 0·8–1·8 cm. long; spatheoles 1·8–2·6 cm. long; peduncles 2–10 mm. long; callus square to cuneate

27. **formosa** (p. 117)

Awns 2–3·4 cm. long; callus cuneate:

Spatheoles 2·2–3·2 cm. long, russet or purplish, enclosing the racemes; peduncles 10–15 mm. long; racemes fairly dense, the internodes 2–2·5 mm. long

28. **schimperi** (p. 118)

Spatheoles 3·5–5 cm. long, grey, the racemes exserted; peduncles 10–50 mm. long; racemes lax, the internodes 2·5–4 mm. long   .   .   .   29. **cyanescens** (p. 120)

Pedicelled spikelets villous:

Pedicel tooth subulate, 0·4–0·8 mm. long or more, rarely less; raceme-base appendage narrowly oblong to linear, mostly 0·5–1·2 mm. long; awns 9–19 per raceme-pair; spatheoles narrow, 4–7 cm. long, the peduncles almost as long or longer; culms slender and wiry

30. **papillipes** (p. 121)

Pedicel tooth broadly triangular, at most 0·2 mm. long:

Basal sheaths pubescent to tomentose, the plant densely caespitose; spatheoles 2·5–5 cm. long; awns up to 2 cm. long, rarely more; pedicelled spikelets muticous or with a short awn-point:

Awns 10–25 per raceme-pair   .   .   .   .   31. **dregeana** (p. 124)

Awns 5–8 per raceme-pair   .   .   .   .   32. **tamba** (p. 126)

Basal sheaths glabrous; awns 4–7 per raceme-pair:

Awns 0·7–1·3 cm. long; callus rounded; spatheoles 1·2–2·3 cm. long; peduncles 3–13 mm. long; pedicelled spikelets muticous or with an awn-point up to 1 mm. long; culms robust and supported by stilt roots   .   .   .   .   33. **umbrosa** (p. 127)

Awns over 1·5 cm. long; callus cuneate; spatheoles 2–4 cm. long; peduncles 10–25 mm. long:

Culms robust, supported by stilt roots; awns 2·2–4 cm. long; pedicelled spikelets usually with a bristle 2–6 mm. long

34. **rudis** (p. 128)

Culms slender, without stilt roots, arising in clumps from a short underground rhizome; awns 1·5–2·5 cm. long; pedicelled spikelets with a short awn point 1–3 mm. long

35. **collina** (p. 130)

IV. Sect. Hyparrhenia

Annuals; panicle loose and leafy; awns 4–10·5 cm. long:

Peduncle bearded on one side near the tip, deflexed at maturity; raceme-base appendage obscure or absent; upper raceme-base 2–2·5 mm. long, subterete, bearing a few scattered hairs, not deflexed; awns 2 per raceme-pair . . . . 36. **madaropoda** (p. 134)

Peduncle uniformly bearded near the tip, more or less straight; raceme-base appendage distinct (except *H. niariensis* var. *macrarrhena*); upper raceme-base 1–1·5 mm. long, flat, densely bearded, deflexed at maturity:

Awns 4–7 per raceme-pair; peduncle hairs white

37. **coleotricha** (p. 136)

Awns 2–3 per raceme-pair; pedicel tooth very short, broadly triangular:

Sessile spikelets 7–11 mm. long; awns 2 per raceme-pair (rarely 3, but then peduncle hairs yellow and sessile spikelets 7·5 mm. long or more):

Bristle of pedicelled spikelet 9–17 mm. long; peduncle hairs white

38. **confinis** (p. 137)

Pedicelled spikelets 11–14 mm. long; sessile spikelets 8–10 mm. long; appendage oblong, 1–2 mm. long; leaf-blades seldom petiolate:

Sessile spikelets glabrescent . 38a. **var. confinis** (p. 137)

Sessile spikelets silky villous . 38b. **var. pellita** (p. 139)

Pedicelleds pikelets 6–10 mm. long; sessile spikelets 7–8·5 mm. long, glabrescent; appendage narrowly oblong, 3–4 mm. long; leaf-blades often petiolate

38c. **var. nudiglumis** (p. 139)

Bristle of pedicelled spikelet 1–10 mm. long; peduncle hairs yellow, or sometimes white (but then bristle of pedicelled spikelet less than 8 mm. long); sessile spikelets 7·5–11 mm. long

39. **niariensis** (p. 140)

Raceme-base appendage oblong to narrowly oblong, 0·5–4 mm. long; lower glume of sessile spikelet rounded, flat, or at most slightly hollowed on the back

39a. **var. niariensis** (p. 140)

Raceme-base appendage absent or very slight; lower glume of sessile spikelet with a broad concave depression along median line . . 39b. **var. macrarrhena** (p. 141)

Sessile spikelets 5–7 mm. long; awns 3 per raceme-pair; peduncle pilose with whitish or pale yellow hairs; pedicelled spikelets 6–8 mm. long, their awn 2–11 mm. long . 40. **welwitschii** (p. 142)

Perennials; awns 2–4 per raceme-pair:

Pedicel tooth obtusely triangular, obscure; sessile spikelets 4–6 mm. long, their awn 1–2·5 cm. long; pedicelled spikelets muticous or with a mucro up to 1 mm. long; panicle narrow, dense . 41. **bracteata** (p. 144)

Pedicel tooth subulate, 0·2–1·5 mm. long; sessile spikelets 6–10 mm. long, their awn 2·2–5·5 cm. long; pedicelled spikelets with a bristle 1–5 mm. long; panicle loose, often scanty   .   .   42. **newtonii** (p. 148)

V. Sect. Arrhenopogonia

Annuals:

Pedicelled spikelets 4–7 mm. long, shortly mucronate

Pedicelled spikelets 10–11 mm. long, with a terminal bristle 11–22 mm. long

Perennials:

Margins of homogamous spikelets without tubercles; lower glume of sessile spikelet hirsute or hispidulous on the back; callus 1 mm. long, acute; awn 2·5–4 cm. long, hispid with hairs 0·3–0·5 mm. long

Margins of homogamous spikelets encrusted with tubercles; lower glume of sessile spikelet with stout prickle hairs on the back; callus 1·5–2 mm. long, pungent; awn 4–5 cm. long, hirtellous with hairs 0·7–1·5 mm. long   .   .   .   .   46. **tuberculata** (p. 155)

VI. Sect. Apogonia

Annuals:

Pedicelled spikelet with an awn 8–20 mm. long; awns 4 per raceme-pair; base of spatheole and supporting ray glabrous, or sometimes bearded at the node   .   .   .   .   .   48a. **var. involucrata** (p. 158)

Pedicelled spikelet with an awn 1–5 mm. long; awns 2(–4) per raceme-pair; base of spatheole and supporting ray pilose

Perennials:

Spikelets awned:

Column of awn with hairs 0·5–1·7 mm. long, the awn 4·5–7·5 cm. long; peduncles 1–3·5 cm. long; spatheoles 3–7 cm. long; pedicelled spikelets with a bristle 2–7 mm. long

Column of awn with hairs 0·2–0·4(–0·5) mm. long, the awn 2–5·5 cm. long:

Homogamous pairs 2 at the base of each raceme; upper raceme-base 1–2 mm. long; racemes 1·5–2(–2·5) cm. long; peduncles 0·3–1·5 cm. long; spatheoles 2–4·5 cm. long   51. **diplandra** (p. 166)

Homogamous pairs 1 at the base of each raceme; upper raceme-base 3–4 mm. long; racemes 2·5–3 cm. long, 6–12-awned per pair

CONSPECTUS OF THE SECTIONS AND SPECIES OF HYPARRHENIA

I. Sect. **Strongylopodia** *W. D. Clayton*, sect. nov.; callo spiculae sessilis fere semicirculari, spiculis glabris, et basibus racemorum inaequalibus plus minusve applanatis, basi longiore 1·5–3 mm. longa distinguenda. Type species: *H. glabriuscula* (Hochst. ex A. Rich.) Anderss. ex Stapf.

Perennials; culms erect or trailing. *Spathate panicle* dense or sparse, with linear-lanceolate spatheoles; peduncles half to longer than the spatheole. *Racemes* paired or sometimes solitary, seldom deflexed, 5–15-awned per pair; raceme-bases more or less flattened, the upper 1·5–3 mm. long, glabrous or softly pilose, sometimes with stiff bristles, unappendaged. *Homogamous pairs* 1 at the base of the lower raceme only. *Sessile spikelets* glabrous; callus very short and rounded, almost semicircular; upper lemma lobes 0·8–1 mm. long; awn glabrous or puberulous. *Pedicelled spikelets* awnless.

It is difficult to decide whether the raceme-bases of *H. glabriuscula* and *H. mobukensis* should be classed as flattened or terete, and hence to determine their affinities. One of the few characters they share in common is the very obtuse callus, which is unlike anything found in other sections (except perhaps for *H. rufa* and *H. cymbaria*). Despite many dissimilarities I feel that they should, on this account, be grouped together, and placed in a separate section.

There is no doubt that these species are correctly placed in *Hyparrhenia*, but the obtuse callus, glabrous awn of *H. mobukensis*, and uncanny resemblance of *H. glabriuscula* to *Cymbopogon* all point to an affinity with the latter genus. It is difficult to resist the inference of proximity to the supposed *Andropogon-Cymbopogon* ancestor of *Hyparrhenia*.

1. **Hyparrhenia glabriuscula** *(Hochst. ex A. Rich.) Anderss. ex Stapf* in Prain, Fl. Trop. Afr. 9: 372 (1918).

Andropogon glabriusculus Hochst. ex A. Rich., Tent. Fl. Abyss. 2: 468 (1851); Hack. in DC., Monogr. Phan. 6: 616 (1889). Type: Ethiopia, *Schimper* 1805 (P, holotype; K, isotype).
Sorghum glabriusculum (Hochst. ex A. Rich.) Kuntze, Rev. Gen. Pl. 2: 791 (1891).
Hyparrhenia amoena ('amaena') Jac.-Fel. in Journ. Agric. Trop. 1: 46 (1954). Type: Senegal, *Berhaut* 1896 (P, holotype).

Caespitose perennial; culms 1·2–2 m. high. *Leaf-sheaths* glabrous; ligule about 1 mm. long; blades up to 30 cm. long and 5 mm. wide, dull green, glabrous or hispidulous above and puberulous beneath, not aromatic. *Spathate panicle* 15–30 cm. long, narrow, congested; spatheoles linear-lanceolate, 2–3 cm. long, green tinged with russet, glabrous; peduncles about half as long as the spatheole, glabrous or rarely with a few hairs towards the top. *Racemes* seldom deflexed, 1·5–2·5 cm. long, 5–7-awned per pair, laterally exserted; raceme-bases unequal, flattened, the upper 2·5–3 mm. long, glabrous or sparsely ciliate on the margins, with a short scarious rim at the tip. *Homogamous spikelets* narrowly lanceolate, 4–6 mm. long, glabrous. *Sessile spikelets* lanceolate, 5 mm. long; callus square to oblong or almost semi-circular, about 0·5 mm. long, broadly rounded; lower glume glabrous, 11-nerved, these sometimes raised, the margins keeled in the upper third and rounded below; awn 1·5–2·5 cm. long, the column puberulous with hairs 0·1 mm. long. *Pedicelled spikelets* narrowly lanceolate, 5 mm. long, glabrous, acuminate; pedicel tooth short, broadly triangular.

ILLUSTRATION. Jacques-Félix in Journ. Agric. Trop. 1: 47, t.6 (1954).

HABITAT. Seasonally swampy soils and river flood-plains.

DISTRIBUTION. West Africa from Senegal to Nigeria; also in Ethiopia, Mozambique and Malawi (Map 2).

SENEGAL. Niokolo Koba (Sept., Oct.), *Adam* 15715, 15811 & 17147; Kidira (Nov.), *Berhaut* 1896.

UPPER VOLTA. Without locality, *Prost.*

GHANA. Han to Tumu (Oct.), *Rose Innes* GC 32427; Naga to Navrongo (Nov.), *Rose Innes* GC 32524; Bawku to Bolgatanga (Oct.), *Rose Innes* GC 31099; Bamboi to Bole (Nov.), *Brand* SLUS 750; Pong Tamale (Oct.), *Rose Innes* GC 32429.

NIGERIA. Biu (Oct.), *de Leeuw* 1314; Buratai (Oct.), *de Leeuw* 1322; Shendam (Nov.), *Clayton* 1465.

ETHIOPIA. Schiré highlands (Oct.), *Schimper* 1805.

MOZAMBIQUE. Angónia (May), *Mendonça* 4203.

MALAWI. Salima, Msekwe (Apr.), *G. Jackson* 478.

H. glabriuscula has a flat glabrous upper raceme-base, and on that account has been placed in sect. *Apogonia* by several authors, despite the unusual length of the raceme-base. A resemblance to *H. gossweileri* is undeniable but, apart from lacking involucral homogamous spikelets, its most characteristic feature, the very blunt callus, is quite atypical of sect. *Apogonia*. A similar rounded callus is found in *H. mobukensis*, whose spikelets are very similar to those of *H. glabriuscula*, and I feel that the two species should be grouped together.

H. glabriuscula is very similar in facies to *Cymbopogon*, but the insertion of the callus is of the typical *Hyparrhenia* type, the lower glume of the sessile spikelet is 2-keeled only at the tip, and the leaves are devoid of aromatic flavour.

2. **Hyparrhenia mobukensis** (*Chiov.*) *Chiov.* in Nuov. Giorn. Bot. Ital. ser. 2, 26: 74 (1919).

Andropogon mobukensis Chiov. in Ann. Bot. Roma 6: 147 (1907). Type: Uganda, *Abruzzi* (TO, holotype).

A. scaettai Robyns in Bull. Jard. Bot. Brux. 8: 223 (1930). Type: Congo Republic, *Scaetta* 158 (BR, holotype).

Hyparrhenia absimilis Pilger in Not. Bot. Gart. Berlin 14: 102 (1938). Type: Tanzania, *Schlieben* 5093 (B, holotype; K. isotype).

Hypogynium absimile (Pilger) Roberty in Boissiera 9: 189 (1960).

Trailing perennial; culms wiry, very slender, up to 1·6 m. long. *Leaf-sheaths* glabrous; ligule 1 mm. long; blades linear-lanceolate, 4–8 cm. long and 2–5 mm. broad, finely acuminate at the tip, rather abruptly narrowed at the base, glabrous, light green, flaccid. *Spathate panicle* of up to 5 distant tiers, each with a single simple ray; spatheoles lanceolate-linear, 4–5 cm. long, light green tinged with purple; peduncles mostly a little longer than the spatheole, sometimes up to twice as long, flexuous, scaberulous or pilose above. *Racemes* paired or rarely solitary, not deflexed, 1·5–3 cm. long, 7–15-awned per pair, terminally exserted; raceme-bases unequal, the upper 1·5–3 mm. long, terete or

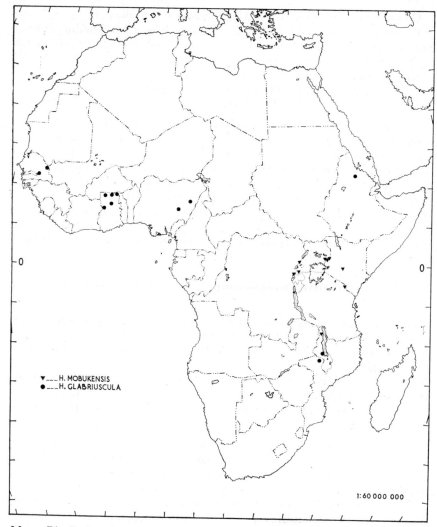

MAP 2. Distribution of *Hyparrhenia mobukensis* and *H. glabriuscula*.

somewhat flattened, softly pilose or sometimes bearing stiff bristles. *Homogamous spikelets* 5–6 mm. long, glabrous. *Sessile spikelets* lanceolate, 4–5 mm. long; callus square or semicircular, 0·5 mm. long, very obtuse; lower glume distinctly 9-nerved, glabrous or scaberulous, laterally keeled in the upper third, broadly rounded below; awn 7–8 mm. long, the column glabrous. *Pedicelled spikelets* lanceolate, 6 mm. long, glabrous, acuminate; pedicel tooth none.

ILLUSTRATION. Fig. 16.

HABITAT. Short grassland in the East African highlands.

DISTRIBUTION. The mountains and highland plateaus of tropical Africa, including Ruwenzori, Elgon, Aberdare, Kilimanjaro and Nyika (Map 2, p. 49).

CONGO REPUBLIC. Muhavura (Mar.), *Scaetta* 158 & 1625.

UGANDA. Ruwenzori, Mobuka (July), *Abruzzi*; Kigezi, Mt. Muhinga (Oct.), *Snowden* 1485; Mt. Elgon (Nov.–Jan.), *Dummer* 3498, *Liebenberg* 1711, *Tothill* 2307, *A. S. Thomas* 2658 & 2664.

KENYA. Kimakia F.R. (Nov.), *Kerfoot* 1353.

TANZANIA. Kilimanjaro, Rongai (Apr.), *Schlieben* 5093.

MALAWI. Nyika Plateau (Jan., Oct.), *E. A. Robinson* 3080, *Robson* 460.

H. mobukensis is readily recognized by its unusual trailing habit and broadly rounded callus, the latter suggesting an affinity with *H. glabriuscula*. The raceme-bases are tantalizingly ambiguous, varying from clearly terete to partially flattened, and from thinly hairy to stiffly bearded. Somewhat similar conditions are found in *H. gazensis*, and the two species may be related.

H. mobukensis is remarkable for two curious features. It has glabrous awns, such as occur in the genera *Andropogon* and *Cymbopogon*, but the callus insertion is of the *Hyparrhenia* type and the leaves are flavourless. It also has a strong tendency to produce solitary racemes; such racemes commence with a normal heterogamous segment.

II. Sect. **Polydistachyophorum** (*Gren. & Godr.*) *W. D. Clayton*, comb. nov.

Andropogon sect. *Polydistachyophorum* Gren. & Godr., Fl. Fr. 3: 469 (1855). Type species: *Andropogon hirtus* L. (=*Hyparrhenia hirta* (L.) Stapf).

Hyparrhenia sect. *Eu-hyparrhenia* Stapf in Prain, Fl. Trop. Afr. 9: 293 (1918), *nom. illegit.*

Hyparrhenia series *Rufae* Stapf, *l. c.* (1918). Type species: *H. rufa* (Nees) Stapf.

Hyparrhenia series *Hirtae* Stapf, *l. c.* (1918). Type species: *H. hirta* (Linn.) Stapf.

Hyparrhenia series *Filipendulae* Stapf, *l. c.* (1918). Type species: *H. filipendula* (Hochst.) Stapf.

Perennials or sometimes annuals; culms usually robust and caespitose, rarely slender or trailing. *Spathate panicle* copious or scanty (often only a single raceme-pair in *H. wombaliensis*); spatheoles narrow, typically rolled around the terminally exserted peduncles at maturity. *Racemes* commonly not deflexed, 2–20-awned per pair, glabrous, rufous or white villous; raceme-bases unequal, the upper terete or filiform, (1·5–)2–5(–10) mm. long, glabrous or sometimes loosely hirsute with soft hairs (stiffly pilose in *H. gazensis*), unappendaged (except *H. andongensis*). *Homogamous pairs* 1 at the base of the lower raceme, and 0–2 at the base of the upper. *Sessile spikelets* 3·5–7(–10) mm. long, the callus very variable but usually short; upper lemma lobes 0·2–0·5 mm. long (up to 1 mm. in species allied to *H. hirta*, but almost absent in *H. wombaliensis*); awn length variable (awns absent in *H. exarmata*), the column puberulous to pubescent (hirtellous in *H. bagirmica* and species allied to *H. filipendula*). *Pedicelled spikelets* narrowly lanceolate, usually awnless but sometimes with an awn up to 10 mm. long; callus none; pedicel tooth variable.

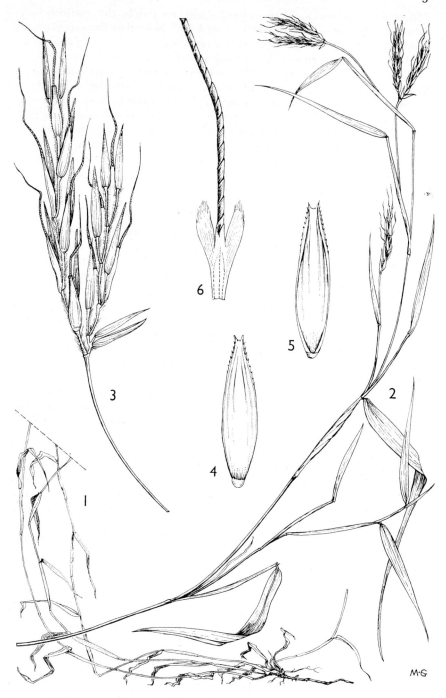

FIG. 16. *Hyparrhenia mobukensis.* **1,** habit, ×⅓; **2,** inflorescence, ×⅔; **3,** raceme-pair, ×3. Sessile spikelet. **4,** lower glume, outer view, ×6; **5,** same, inner view, ×6; **6,** upper lemma and awn, ×6. **1** drawn from *O. Kerfoot* 1353; **2–6** from *E. A. Robinson* 3080.

Most of the species can be recognized readily enough by the relatively long, glabrous, terete or filiform upper raceme-base. (Note that the raceme-bases are always pubescent in the fork, and the term glabrous is taken to refer to the shank only.) The raceme-base is sometimes softly hirsute or adorned with a few bristles, but only in *H. gazensis* does the luxuriance of its indumentum approach that of sect. *Pogonopodia*. The raceme-base may be rather short, though still definitely terete, in *H. quarrei*, *H. gazensis* and *H. finitima*, and some care is needed to distinguish these species from sect. *Pogonopodia*, with which the present section tends to intergrade. Other typical, but inconstant, features of the section are many-awned, non-deflexed racemes terminally exserted from the narrow spatheoles, and a short obtuse or acute sessile spikelet callus.

H. andongensis, with its appendaged raceme-base, is something of an oddity, but its long filiform upper base seems a clear indication that it should be included here.

As will be seen, sect. *Pogonopodia* is treated as a continuum of evenly spaced, closely related species of approximately equal value. The present section also is best regarded as a continuum, but additionally complicated by great variation in the size and distinctness of its noda (or species). They can be roughly grouped into four classes as follows:

1. Major species; a few common species of wide distribution which dominate the section. They are rather variable, but in a successful taxon of great geographical amplitude this is not inconsistent with the rank of species.

2. Satellite species; secondary noda at the periphery of a major species. Whether they should be separated from the major species or not is largely a matter of subjective opinion.

3. Minor species; well-marked secondary noda which can usually be identified as species without great difficulty. However the allocation of specific rank is to some extent dependent upon the circumscription of the major species, for it would not always be compatible with the interpretation of the latter in a very wide sense.

4. Linking species; intermediate populations which link together the major species (and are sometimes contingent upon minor ones also), without obvious discontinuities beyond a significant shift in character frequencies. They are very variable and I suspect that each contains a miscellany of less successful introgression products maintained by apomixy.

It is the treatment of what I have called linking species that constitutes the greatest problem in the section. The major species are too distinct for the occurrence of intermediates to be taken as license for lumping, and so there remain two possibilities:

(a) The major species are circumscribed so widely as to apportion the intermediates between them. The resulting classification has the appeal of simplicity. But the simplicity is deceptive, for it conceals considerable heterogeneity within the species, and is often an excuse for uncritical boundaries between them. Furthermore widely conceived major species would tend to engulf many of the satellite and minor species, thus adding to the heterogeneity of the former.

(b) The circumscription of the major species is restricted to the nodal populations, and the connecting clusters are separated as distinct species. Such an approach reveals more of the structure and interrelationships within the section, but at the expense of a more complex taxonomy in which the naming of individual specimens is more troublesome.

It seems to me that a classification should not be judged solely upon its ready comprehensibility—a field manual is the place for simplifications if such are desirable in a particular context—but should reflect the situation found in the field, thereby suggesting relationships whose nature might subsequently be tested by experimental methods. I have therefore adopted the second alternative.

Stapf (1918) divided the section into 3 series each dominated by a major species. These series are too ill-defined to be of much taxonomic value, and I have not adopted them formally, but they provide a useful introduction to the structure of the section. The main characters, greatly simplified, are set out in Table 5, and the distribution of the species within the series is shown in Table 6. The key to the taxonomy of this section lies in a satisfactory identification of the major noda, for it is only after these species have been contained that the classification of the residue becomes a feasible proposition. The matter is discussed further under the species concerned.

TABLE 5. Principal characters of species groups in sect. *Polydistachyophorum*

	Series *Rufae*	Series *Hirtae*	Series *Filipendulae*
Homogamous pairs on upper raceme	0–1	0–1	2
Awns per raceme-pair	5–15	5–15	2–3
Upper raceme-base	short	short	long
Callus	obtuse to acute	acute to pungent	pungent
Spikelet hairs	rufous	white	glabrous or white
Pedicelled spikelets	awnless	awnless	awned
Awn hairs	short	short	long

TABLE 6. The species groups of sect. *Polydistachyophorum*

	Series *Rufae*	Series *Hirtae*	Series *Filipendulae*	
Major	H. rufa H. nyassae	H. hirta	H. filipendula	
Satellite	H. smithiana H. exarmata H. dichroa	H. violascens	H. figariana	
Minor	H. gazensis H. finitima	H. bagirmica H. wombaliensis	H. familiaris H. barteri	H. andongensis
Linking	H. poecilotricha	H. anamesa H. quarrei H. griffithii		

3. **Hyparrhenia nyassae** (*Rendle*) *Stapf* in Prain, Fl. Trop. Afr. 9: 313 (1918).

Andropogon nyassae Rendle in Journ. Bot. 31: 358 (1893). Type: Malawi, *Buchanan* 1423 (K, isotype).

A. rufus var. *auricomus* Pilger in Engl., Bot. Jahrb. 30: 268 (1901). Type: Tanzania, *Goetze* 901 (K, isotype).

Cymbopogon chrysargyreus Stapf in Journ. de Bot. sér 2, 2: 213 (1909). Type: Central African Republic, *Chevalier* 5366 (P, holotype; K, isotype).

C. solutus forma *trichophyllus* Stapf, *l. c.* (1909). Type: Central African Republic, *Chevalier* 5406 (P, syntype; K, isosyntype), *Chevalier* 15407 (P, syntype).

Andropogon chrysargyreus (Stapf) Stapf ex A. Chev., Sudania 1: 77 (1911).

A. lasiobasis Pilger in Fries, Wiss. Ergebn. Schwed. Rhod.-Kongo Exped. 1911–12, 1: 197 (1916). Type: Zambia, *Fries* 1089 (UPS, holotype).

Cymbopogon nyassae (Rendle) Pilger in Engl., Bot. Jahrb. 54: 287 (1917).

Andropogon lugugaënsis Vanderyst in Bull. Agric. Congo Belge 9: 241 (1918), *nom. prov.* Type: Congo Republic, *Vanderyst* 6073 (BR, holotype).

Hyparrhenia chrysargyrea (Stapf) Stapf in Prain, Fl. Trop. Afr. 9: 312 (1918).

H. vulpina Stapf, *l. c.*: 310 (1918). Type: Angola, *Gossweiler* 3120 (K, lectotype).

Cymbopogon vanderystii De Wild. in Bull. Jard. Bot. Brux. 6: 24 (1919). Type: Congo Republic, *Vanderyst* 373 (BR, holotype).

Andropogon vanderystii De Wild., *l. c.* (1919), *in synon.*

Hyparrhenia vanderystii (De Wild.) Vanderyst in Bull. Agric. Congo Belge 11: 144 (1920).

H. schmidiana A. Camus in Journ. Agric. Trop. 2: 201 (1955). Type: Vietnam, *Schmid* 2460 (P, holotype).

Cymbopogon schmidianus (A. Camus) A. Camus ex Schmid, Fl. Agrost. Indochine: 229 (1959), sine bibl. cit.

Caespitose perennial; culms 0·5–1·5 m. high and 1–4 mm. in diameter at the base. *Leaf-sheaths* at base of plant pubescent to tomentose with white hairs; ligule up to 1·5 mm. long; blades rigid, up to 45 cm. long and 2–5 mm. wide, often coarsely hairy near the base. *Spathate panicle* lax, 15–45 cm. long; spatheoles linear, 3–6 cm. long, russet, at length rolled; peduncles usually longer than the spatheole, with spreading white hairs above. *Racemes* tardily and imperfectly deflexed, 2–3 cm. long, 8–13-awned per pair, fulvous, or greyish-yellow with the callus, internodes and pedicels white, terminally exserted on the flexuous peduncle; raceme-bases unequal, the upper 2–3 mm. long, glabrous or sometimes sparsely pilose, the articulation with the peduncle tip marked by a ring of white hairs, these sometimes brown or absent. *Homogamous pairs* 1 at the base of the lower raceme; spikelets narrowly lanceolate, 5–6 mm. long, acute. *Sessile spikelets* narrowly lanceolate, 5–6 mm. long; callus linear to slenderly cuneate, 0·8–1·2 mm. long, acute to very narrowly truncate; lower glume yellowish green to violet, typically densely pubescent with golden yellow hairs; awn 2–4 cm. long, the column fulvously pubescent with hairs 0·2–0·5 mm. long. *Pedicelled spikelets* linear-lanceolate, 4·5–7 mm. long, acute; pedicel tooth subulate, 0·2–0·5 mm. long.

MINOR VARIATIONS. The basal sheaths, though usually almost tomentose, are occasionally only sparsely pubescent. A few specimens with glabrous, apparently annual, bases from Jebel Marra, Sudan (*Blair* 163, 174) seem also to belong here, for the copious rufous indumentum of the spikelets accords ill with any other species.

There is sometimes a homogamous pair at the base of the upper raceme as well as the lower. Very rarely the sessile spikelets are almost glabrous (*G. Jackson* 1096, from Kotakota, Malawi). *Bosser* 17549 from Ankaratra, Madagascar should apparently be referred to this species, for the basal sheaths are sparsely pilose and some of the spikelets bear pale yellow hairs (others are white hairy). The raceme structure, however, resembles *H. griffithii*.

The spikelets are often infested with the smut *Sphacelotheca barcinonensis* Riofrio.

CHROMOSOME NUMBER. Uganda, $2n = 20$ (Tateoka, 1965a & b, as *H. rufa*)!. Congo Republic, $2n = 40$ (Celarier & Harlan, 1957).

HABITAT. Savanna woodland, particularly moist places and swamp edges.

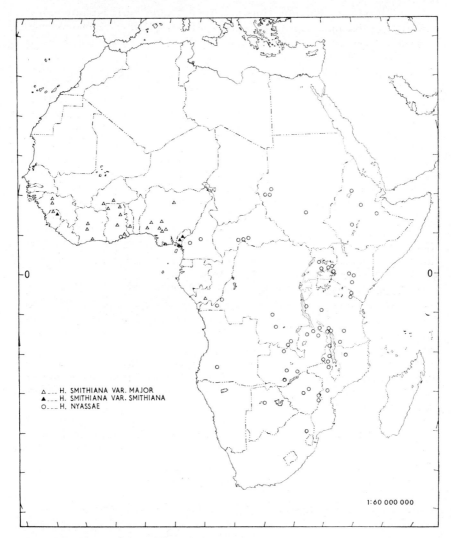

MAP 3. Distribution of *Hyparrhenia nyassae* and *H. smithiana* var. *smithiana* and var. *major*.

DISTRIBUTION. Tropical Africa from Cameroun and Sudan to Transvaal; only one specimen has been recorded from west of the Cameroun mountains. Also found in Thailand and Vietnam (Map 3).

THAILAND. Chiengmai (Dec.), *Smitinand* 16797.
VIETNAM. Haut-Donnai, Juga, *Schmid* 2460 (P).

GHANA. Accra plains, Doyum (Nov.), *Ankrah* GC 20109.

CAMEROUN. Nanga-Eboko (Nov.), *Vaillant* 2303; Sabal Maba, Minim (Sept.), *Letouzey* 5922, 5935; Meiganga to Ngaundere (Oct.), *Koechlin* 7220; Dzerkoka (Dec.), *Raynal* 12355.

CENTRAL AFRICAN REPUBLIC. Krébedjé, Fort Sibut (Sept.), *Chevalier* 5406, 5407; Confluence of Kémo R. (Sept.), *Chevalier* 5366; Bambari (Oct.), *Audru & Boudet* 3429, 3430.

CONGO REPUBLIC. Kisantu, Horneni (Apr.), *Callens* 1301; Kisantu (May), *Vanderyst* 373; Kamina (Apr.), *Quarré* 3021; Mérode, *Vanderyst* 23069; Nlemfu, Luguga R. (Apr.), *Vanderyst* 6073; Elisabethville (May), *Quarré* 324; Koravia, *Quarré* 1943.

SUDAN. Jebel Marra (Jan.), *J. K. Jackson* 2635; Saur (Jan.), *Wickens* 979; Tura Tonga (Feb.), *Blair* 255; Gollol (Jan., June), *Blair* 49, 204.

ETHIOPIA. Gondar (Aug., Sept.), *Chiovenda* 1581, 2480; Mulu-Saiu (May), *Mooney* 4782, 4783; Nadda (Oct.), *Mooney* 6219; Harar (Sept.), *Burger* 933.

UGANDA. Nakasongola (Apr.), *Langdale-Brown* 2082; Kakoge (May), *Tateoka* 3230; Bushenyi (May), *Snowden* 1363; Kagoye, Buruli (May), *A. S. Thomas* 3756; Namasagali (June), *G. H. S. Wood* 956; Tororo (Mar.), *Michelmore* 1197.

KENYA. Bungoma (June), *Bogdan* 4060; Embu (May), *Schantz* 994; Thika to Fort Hall (Mar.), *Bogdan* 1547; Thika (Jan.), *Bogdan* 2181; Nairobi (June), *Linton* 151.

TANZANIA. Marungu (Jan.), *Schlieben* 4635; Tabora, *Speke & Grant*; Iringa, Njombe, *Emson* 127; Ufipa, Nsangu Mt. (Mar.), *McCallum Webster* T88; Mbeya, Chalanga (Feb.), *Vesey-FitzGerald* 2272; Kinga Mts., *Goetze* 901; Songea (Feb.), *Milne-Redhead & Taylor* 8830.

MOZAMBIQUE. Marrupa (Feb.), *Torre and Paiva* 10677; Vila Coutinho (May), *Mendonça* 4147; Gorongose (Nov.), *Torre & Paiva* 9170.

MALAWI. Without locality, *Buchanan* 1423; Mzimba (Mar.), *G. Jackson* 438; Kotakota, Soni (Feb.), *G. Jackson* 1096; Ft. Manning, Nyoka (Apr.), *G. Jackson* 617; Lilongwe (Apr.), *G. Jackson* 459.

ZAMBIA. Abercorn to Kasama (Apr.), *Vesey-FitzGerald* 1648; Luwingu (Oct.), *Fries* 1089; Mbesuma (Oct.), *Astle* 968; Solwezi (June), *Milne-Redhead* 490; Lusaka (May), *Trapnell* 2019; Mumbwa (June), *Trapnell* 2082; Choma (Apr.), *Astle* 2999.

RHODESIA. Hartley (Apr.), *Wild* 4547; Ingesi, *Hobbs* 677; Marandellas, Digglefold (Feb.), *Corby* 4; Umtali (Mar.), *Chase* 6371; Melsetter, Sibu (Dec.), *Crook* 421; Shangani (Mar.), *Goldsmith* 40/56.

BOTSWANA. Maun (June), *van Son* 28619.

ANGOLA. Munongue (Apr.), *Gossweiler* 3120; Nova Lisboa (May), *Gossweiler* 10746.

SOUTH AFRICA. Transvaal: Machadodorp (Mar.), *Pole-Evans* 3682; Schoemanskloof (Feb.), *de Winter & Codd* 432.

H. nyassae can usually be identified readily enough by the felt of white hairs on its basal sheaths, and it is further distinguished from *H. rufa* by its longer pointed callus and generally hairier racemes. The species grades imperceptibly into *H. quarrei*, which is distinguished by its white haired racemes, and also into *H. griffithii* (*q.v.*).

A distinction has been made between low, narrow leaved, tussocky specimens (*H. nyassae*), and tall specimens with stout leafy culms and broader blades

(*H. vulpina*). However it transpires that variation between these extremes is quite continuous, and the distinction is untenable; it is probably merely a response to environmental conditions. The tall form has also been called *H. chrysargyrea*, a name which has more often been misapplied to *H. smithiana*. Although *H. chrysargyrea* was described as an annual, the holotype at Paris is an annual-like piece broken from a perennial clump (the isotype at Kew lacks a base). The basal sheaths are glabrous, a condition not entirely unexpected in a fragmentary and well-worn specimen. Other specimens from the Central African Republic are undoubtedly *H. nyassae*, and I have no hesitation in consigning *H. chrysargyrea* to the synonymy of this species.

4. **Hyparrhenia smithiana** (*Hook. f.*) *Stapf* in Prain, Fl. Trop. Afr. 9: 314 (1918).

Andropogon smithianus Hook. f. in J. Linn. Soc., Bot. 7: 232 (1864); Hack. in DC., Monogr. Phan. 6: 622 (1889). Type: Cameroon Mtn., *Mann* 1342, 2079 (both K, syntypes).
Sorghum smithianum (Hook. f.) Kuntze, Rev. Gen. Pl. 2: 792 (1891).

4a. var. **smithiana**

Perennial; culms 40–100 cm. high and 2–3 mm. in diameter at the base; basal parts with an indumentum of purplish-red hairs, but these often very scanty and pallid. *Leaf-blades* 2–5 mm. wide. *Racemes* not, or imperfectly, deflexed, with a very dense indumentum of dark rufous or chocolate coloured hairs. Otherwise similar to var. *major* (below).

CHROMOSOME NUMBER. Cameroon Mt., 2n = 40 (Vohra, 1966).

HABITAT. A constituent of high altitude grasslands: Cameroon Mt. 2000–3000 m., Bamenda 1400–2500 m., Bintumane 1400–1700 m.

DISTRIBUTION. The mountains of Cameroun and Sierra Leone (Map 3, p. 55).

SIERRA LEONE. Mt. Loma (July), *Jaeger* 6975; Bintumane Peak (Jan., July, Nov.), *T. S. Jones* 116, *Bakshi* 289, *Jaeger* 322, *J. K. Morton* SL 2669.
CAMEROUN. Bambui (Dec.), *Boughey* GC 10752; Lake Oku (Jan.), *Keay & Lightbody* FHI 28499; Bafut-Ngemba F.R., *Hepper* 2848; Bamenda (Jan.), *Migeod* 343; Cameroon Mt. (Dec.–Feb.), *Mann* 1342, 2079, *Boughey* GC 12564, *Maitland* 1042, *J. K. Morton* K824, *Steele* 51.

4b. var. **major** *W. D. Clayton*, var. nov.; affinis varietati typicae, sed culmis robustioribus et spiculis pallidioribus differt. Typus: Ghana, *Rose Innes* GC 32385 (K, holotype).

Hyparrhenia chrysargyrea auct., Fl. W. Trop. Afr., ed. 1, 2: 591 (1936), *non* Stapf.

Perennial; culms 1·5–2·5 m. high and 3–5 mm. in diameter at the base. *Leaf-sheaths* at base of plant clothed in a felt of purplish-red hairs; ligule up to 3 mm. long; blades rigid, up to 75 cm. long or more, 4–6 mm. wide, often coarsely bearded above the ligule, scabrid on the margins. *Spathate panicle* lax, 30–90 cm. long; spatheoles linear-lanceolate. 3·5–7 cm. long, reddish, at length tightly rolled; peduncles about as long as spatheoles, with spreading hairs towards the top. *Racemes* at length deflexed, 2·5–3 cm. long, 8–14-awned per pair, fulvous to rufous or with the callus beards and lower hairs of the internodes and pedicels white, subterminally exserted on the curved peduncle; raceme-bases unequal, the upper 2–3 mm. long, glabrous, less often pubescent, and rarely with a few white bristles, their articulation with the peduncle tip marked by a corona of rufous hairs. *Homogamous pairs* 1 at the base of the lower raceme; spikelets narrowly lanceolate, 5–6 mm. long, acute. *Sessile spikelets* narrowly lanceolate, 4·5–6 mm. long; callus cuneate, 0·4–0·8(–1) mm. long, obtuse to narrowly truncate; lower glume purplish, not shiny, usually concealed by silky golden hairs; awn 2–3·5 cm. long, the column fulvously pubescent with hairs 0·2 mm. long. *Pedicelled spikelets* narrowly lanceolate, 5–6 mm. long, acute; pedicel tooth subulate, 0·1–0·5 mm. long.

MINOR VARIATIONS. The indumentum of the sessile spikelets varies considerably from densely villous to almost glabrous (*Ankrah* GC 20228 from Ghana), and from dark rufous to pallidly fulvous or even almost white. The spikelets themselves may be much smaller than usual (only 3·5 mm. long in *Ankrah* GC 20228). Very rarely there is a homogamous pair at the base of the upper raceme also.

The spikelets are frequently attacked by the smut fungus *Sphacelotheca barcinonensis* Riofrio.

CHROMOSOME NUMBER. Ghana, 2n = *c.* 40 *Rose Innes*, GC 32487!; 2n = *c.* 20 *Rose Innes*, GC 32485!. Regrettably, root tip preparations fell short of the perfection required to confirm the precise chromosome number, but at least the ploidy level is clear.

HABITAT. A tall savanna grass of lowland regions. The reddish felted base is often a conspicuous feature of the ground layer during the dry season.

DISTRIBUTION. The Guinea savanna of West Africa (Map 3, p. 55).

UPPER VOLTA. Samandéni (Oct.), *Kmoch* 164; Diebougou to Ouassa (Oct.), *Rose Innes* GC 31514; Leo to Po (Oct.), *Rose Innes* GC 31075.

GUINEA. Timbi Madina (Jan.), *Langdale Brown* 2555; Dalaba (Oct.), *Adam* 12672.

SIERRA LEONE. Musaia (Dec.), *Deighton* 4522; Tingi Mts. (Apr., Dec.), *Morton & Gledhill* SL 1945, SL 3033; Loma Mt. (Nov.) *J. K. Morton* SL 2558.

IVORY COAST. Séguéla (Oct.), *Adjanohoun* 288a, 319a; Zuénoula to Vavoua (Oct.), *Adjanohoun* 382a; Asakra (Aug.), *Boughey* GC 18628; Dabou (Jan.), *Adjanohoun & Ake Assi*.

GHANA. Wa (Oct.), *Rose Innes* GC 32385; Bamboi to Bole (Oct.), *Rose Innes* GC 30646; Salaga to Tamale (Oct.), *Rose Innes* GC 32440; Gambaga (Oct.), *Ankrah* GC 20271; Atebubu (Nov.), *Clayton* 387; Kpong (May), *Rose Innes* GC 30364.

TOGO. Badou to Atakpame (Sept.), *Rose Innes* GC 31349.

NIGERIA. Zaria (Nov.), *Thatcher* S. 529; Lokoja (Oct.), *Dalziel* 293; Ayere (Dec.), *Clayton* 616; Olokemeji (Nov.), *Clayton* 567; Ibadan (May), *Townrow* 22; Ekpeshi (Nov.), *Keay* FHI 28083.

CAMEROUN. Vokré (Oct.), *Koechlin* 7308.

REPUBLIC OF CONGO. Nguedi (June), *Koechlin* 2696.

H. smithiana is readily recognized by the remarkable felt of dark hairs about its basal sheaths. In the absence of the base it is not always easy to distinguish from *H. rufa*, for the two species overlap to some extent. The most useful discriminatory characters are then the ring of rufous hairs at the foot of the raceme-bases, the longer sessile spikelet and callus, laxer racemes, and generally softer and denser spikelet indumentum.

At one time two species were recognized—a short tussock grass on Cameroon Mt. (*H. smithiana*) and a stemmy lowland species (misnamed *H. chrysargyrea*) —but the subsequent collection of specimens from Bamenda and Bintumane has bridged the difference in stature between the two species (Fig. 17,). Nevertheless the noticeably darker spikelet indumentum of the highland plants is an additional character providing tenuous support for their continued separation at varietal level. I was unable to find any reliable characters for separating the two ploidy levels indicated by chromosome counts.

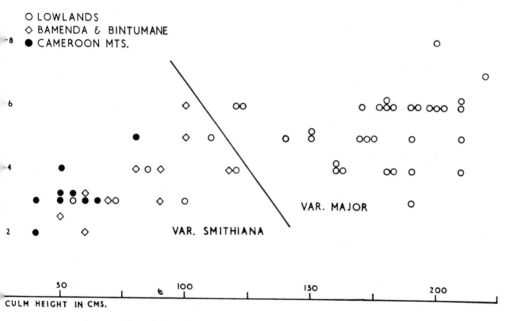

O LOWLANDS
◇ BAMENDA & BINTUMANE
● CAMEROON MTS.

VAR. MAJOR

VAR. SMITHIANA

CULM HEIGHT IN CMS.

FIG. 17. *Hyparrhenia smithiana*. Leaf width plotted against culm height.

H. smithiana is very closely allied to *H. nyassae.*, being distinguished by the dark hairs of its basal sheaths and the shorter more obtuse callus of its sessile spikelets. However, the callus characters overlap and become ambiguous in the intermediate range, so that if the basal parts are missing the species often cannot be distinguished. *H. smithiana* also tends to have taller culms and broader leaf-blades, but there is so much overlap that these characters are useless for discrimination. We may also note the parallel variation in colouring and stature within the two species. It seems clear that *H. smithiana* and *H. nyassae* are close relatives separated by a mountian barrier in Cameroun (*H. subplumosa* and *H. diplandra* are a similar pair of species). It is arguable whether species is the correct rank for such entities, but I feel that the introduction of a hierarchy of infra-specific categories in *Hyparrhenia* would generate more difficulties than it solves.

5. **Hyparrhenia exarmata** *(Stapf)* Stapf in Prain, Fl. Trop. Afr. 9: 308 (1918).

Cymbopogon exarmatus Stapf in Journ. de Bot., sér. 2, 2: 210 (1909). Type: Central African Republic, *Chevalier* 10509 (K, holotype).

Andropogon rufus var. *exarmatus* (Stapf) Stapf ex A. Chev., Sudania 1: 180 (1911).

Perennial, or perhaps sometimes annual; culms 1·5 m. high, erect. *Leaf-sheaths* glabrous; ligule 1–2 mm. long; blades up to 60 cm. long and 9 mm. wide, glabrous or loosely pilose. *Spathate panicle* narrow, 20–40 cm. long; spatheoles linear-lanceolate, 3–6 cm. long, reddish, tightly rolled about the peduncle at maturity; peduncles about as long as the spatheole, minutely pubescent to pilose above. *Racemes* not deflexed, 1·5–2·5 cm. long, with 12–19 spikelets per pair, terminally exserted, rufous; raceme-bases unequal, the upper 3 mm. long, glabrous. *Homogamous pairs* 1 at the base of the lower raceme only; spikelets 3·5–5 mm. long, similar to the pedicelled spikelets. *Sessile spikelets* lanceolate, 3·5–4 mm. long; callus short (0·25 mm.), obtuse; lower glume reddish brown, glossy, with a sparse indumentum of rufous hairs; lower lemma lanceolate, 3 mm. long, hyaline, nerveless, loosely ciliate on the margins; upper lemma narrowly ovate, 3 mm. long, 3-nerved, minutely truncate and mucronate at the apex, sparsely ciliate on the margins. *Pedicelled spikelets* narrowly lanceolate, 3 mm. long, acute; pedicel tooth triangular, up to 0·2 mm. long.

MINOR VARIATIONS. Some of the spikelets in a raceme (3–10 per pair), may bear awns 5–10 mm. long; the Kew specimen of *Troupin* 2007 (type number of *H. parvispiculata*) is just such an intermediate specimen.

HABITAT. Savanna, often bordering swampy soils.

DISTRIBUTION. An uncommon species scattered through the Sudan savanna belt (Map. 4).

UPPER VOLTA. Bobo-Dioulasso (Oct.), *Kmoch* 120.
CHAD. Bahr Salamat (June), *Chevalier* 10509.
CENTRAL AFRICAN REPUBLIC. Ouaka (Oct.), *Tisserant* 2070; Bambari (Oct.), *Tisserant* 432.
CONGO REPUBLIC. Libenge to Zongo (Nov.), *Lebrun* 1647.
SUDAN. Tambura (July), *Myers* 7049.

A curious species distinguished by its awnless racemes. It is clearly a peripheral fragment from the polymorphic *H. rufa* complex but, apart from the lack of awns, it is also distinguished from that species by the greater number of spikelets in the raceme, so that separation of the two species seems to be justified. It should not be confused with specimens of *H. rufa* in which the awns have been suppressed by smut attack, a condition readily apparent from all but the most superficial glance.

6. **Hyparrhenia rufa** *(Nees)* Stapf in Prain, Fl. Trop. Afr. 9: 304 (1918).

Trachypogon rufus Nees, Agrost. Bras.: 345 (1829). Type: Brazil, *Martius* (M, holotype).

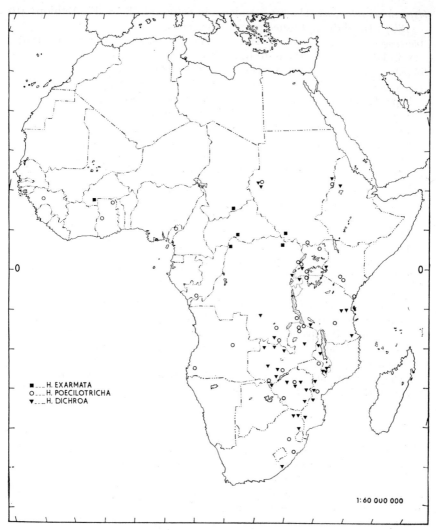

MAP 4. Distribution of *Hyparrhenia exarmata*, *H. poecilotricha* and *H. dichroa*.

Andropogon xanthoblepharis Trin. in Mem. Acad. Sci. Petersb. sér. 6, 2: 281
    (1832), & Sp. Gram. Icon. t. 330 (1836); Hack. in DC., Monogr. Phan.
    6: 637 (1889). Type: Republic of Congo, probably *Smith* (K, isotype); a
    dissected spikelet at LE must presumably be taken as holotype.

A. rufus (Nees) Kunth, Enum. Pl. 1: 492 (1833); Hack. in DC., Monogr. Phan.
    6: 621 (1889); Stapf in Thistleton-Dyer, Fl. Cap. 7: 358 (1898).

A. altissimus Hochst. ex A. Braun in Flora 24: 277 (1841), *non* Raspail (1825),
    nec Colla (1836). Type: Ethiopia, plants cultivated at Karlsruhe from seed
    collected by Schimper (K, isotype).

A. fulvicomus Hochst. in sched., Schimp., Iter Abyss. 2: 1118 (1842); A. Rich.,
    Tent. Fl. Abyss. 2: 463 (1851). Type: Ethiopia, *Schimper* 1118 (K, isotype).

A. fulvicomus var. *approximatus* Hochst. in sched., Schimp., Iter Abyss. 2: 928
    (1842). Type: Ethiopia, *Schimper* 928 (K, isotype).

5

Hyparrhenia fulvicoma (Hochst.) Anderss. in Schweinf., Beitr. Fl. Aethiop: 310 (1867). Invalidly published prior to the genus.

Andropogon rufus var. *fulvicomus* (Hochst.) Hack. in Bol. Soc. Brot. 5: 213 (1887), & in DC., Monogr. Phan. 6: 621 (1889).

Sorghum rufum (Nees) Kuntze, Rev. Gen. Pl. 2: 792 (1891).

Andropogon bouangensis Franch. in Bull. Soc. Hist. Nat. Autun 8: 333 (1895). Type: Republic of Congo, *Thollon* 1078 (P, holotype; K, isotype).

Cymbopogon rufus (Nees) Rendle, Cat. Afr. Pl. Welw. 2: 155 (1899).

C. rufus var. *fulvicomus* (Hochst.) Rendle, *l. c.* (1899).

C. rufus var. *major* Rendle, *l. c.* (1899). Type: Angola, *Welwitsch* 7409 (K, isotype).

Andropogon rufus var. *glabrescens* Chiov. in Ann. Istit. Bot. Roma 8: 288 (1903). Type: Eritrea, *Pappi* 1242 (FI, lectotype).

A. yinduensis Vanderyst in Bull. Agric. Congo Belge 9: 243 (1918), *nom. prov.*, & *op. cit.* 11: 144 (1920), *in synon.*

Hyparrhenia altissima Stapf in Prain, Fl. Trop. Afr. 9: 307 (1918). Based on *Andropogon altissimus* Hochst. ex A. Braun.

H. rufa var. *major* (Rendle) Stapf in Prain, Fl. Trop. Afr. 9: 306 (1918).

H. rufa var. *fulvicoma* (Hochst.) Chiov. in Nuov. Giorn. Bot. Ital. 26: 74 (1919).

H. hirta var. *brachypoda* Chiov. in Atti R. Acad. Ital., Mem. Cl. Sci. Fis. 11: 63 (1940). Type: Ethiopia, *Piovano* 79 (FI, holotype).

H. parvispiculata Bamps in Bull. Jard. Bot. Brux. 25: 391 (1955). Type: Congo Republic, *Troupin* 2007 (BR, holotype). [The specimen at K resembles *H. exarmata*.]

H. vulpina subsp. *longipes* A. Camus in Bull. Soc. Bot. Fr. 107: 207 (1960). Type: Madagascar, *Humbert* 29669 (P, holotype).

6a. var. **rufa**

Perennial, or sometimes annual; culms 30 cm. to 2·5 m. high, tufted, erect. *Leaf-sheaths* glabrous; ligule up to 2 mm. long; blades rigid, 30–60 cm. long and 2–8 mm. wide. *Spathate panicle* lax or contracted and fasciculate, in robust specimens branching to the third degree, in meagre specimens much reduced; spatheoles linear-lanceolate, 3–5 cm. long and 1–3 mm. wide in profile, at length reddish and rolled around the peduncle; peduncles sometimes shorter than the spatheole but usually exserted, often by 2 cm. or more, glabrous or white pilose above. *Racemes* not deflexed (or very rarely so), (1·5–)2–2·5 cm. long, (7–)9–14-awned per pair, fulvous to rufous or with the callus beards and lower part of the internodes and pedicels white, terminally exserted; raceme-bases unequal, the lower 1–2 mm. long, the upper 2–3·5(–4) mm. long, glabrous (very rarely with a few long hairs), occasionally both bases more or less connate, beardless at the foot. *Homogamous pairs* 1 at the base of the lower, or of both, racemes; spikelets similar to the pedicelled. *Sessile spikelets* lanceolate, (3–)3·5–4·5(–5) mm. long; callus 0·2–0·8 mm. long, short and rounded to cuneate and narrowly truncate; lower glume yellowish brown to reddish brown often tinged with violet, rarely green or glaucous, usually glossy, glabrous to pubescent but typically with a scanty covering of stiff rufous hairs; awn 2–3 cm. long, the column rufously pubescent. *Pedicelled spikelets* narrowly lanceolate, 3–5 mm long, acute or rarely mucronate; pedicel tooth triangular, up to 0·3 mm. long.

ILLUSTRATIONS. Trin., Sp. Gram. Icon. t. 330 (1836); C. E. Hubbard, E. Afr. Pasture Pl. 2: 12, t. 33(1927); Robyns, Fl. Agrost. Congo Belge 1: 167, t. 13 (1929); A. Chevalier in Rev. Bot. Appliq. 13: 873, t. 25 (1933); Hutchinson & Dalziel, Fl. W. Trop. Afr. ed. 1, 2: 590, t. 377 (1936); Hubbard & Vaughan, Grasses Mauritius & Rodriguez: 112, t. 14 (1940); Robyns, Fl. Sperm. Parc Nat. Albert 3: 63, t. 7 (1955); Andrews, Fl. Pl. Sudan 3: 467, t. 117 (1956).

MINOR VARIATIONS. A most variable species, the following being the chief sources of variations:

1. It is often difficult to establish whether plants are annual or perennial from herbarium specimens, as these commonly consist of pieces broken from a larger clump, but at least a few of the specimens seen were undoubtedly annual.

2. Stature varies from 0·3 m. to 2·5 m., the dwarfed specimens being associated with trampled or grazed areas.

3. The panicle varies from dense and fastigiate to lax and spreading.

4. The spatheole varies from narrowly lanceolate, enclosing the racemes, to linear and rolled around the peduncle of the exserted racemes. It varies somewhat with age, and it is common to find the distal racemes exserted, while the proximal ones are enclosed.

5. The peduncle may be glabrous, pubescent, or stiffly villous with long white hairs.

6. The length of the raceme-bases is variable, and they may be more or less connate. They are occasionally hairy; the longer base and more numerous spikelets per raceme-pair distinguish such plants from H. dichroa.

7. A homogamous pair may, or may not, occur at the base of the upper raceme; exceptionally there are two pairs. The number is not always constant in the same panicle. Very rarely one or both of the racemes are branched.

8. Spikelet indumentum varies from almost glabrous to almost villous, and from off-white to chocolate brown or orange. The white-haired spikelets are otherwise identical to normal H. rufa, and are regarded as the extreme expression of a tendency to pallid hairs which can be discerned in the latter.

9. Spikelet and callus lengths cover a wide range, occasionally exceeding 5 mm. In general the longer spikelets also have a longer callus.

10. The lower glume itself is usually yellowish or brownish, but is often violet tinged and may be olive green or glaucous.

Detailed scoring of 370 African specimens revealed no obvious discontinuities in the variation, nor any clear correlation between characters, though certain generalities emerged. Thus glabrous or pubescent peduncles are about twice as frequent as villous ones; while specimens with a homogamous pair on the upper raceme are about half as common as those without, and tend to occur more frequently in eastern Africa. It is concluded that the species and varieties which have been based upon one or other of these characters are artificial devices devoid of biological meaning and impossible to segregate in practice.

The spikelets are frequently attacked by the smut fungi Sphacelotheca barcinonensis Riofrio and Sorosporium zundelianum Ciferri.

CHROMOSOME NUMBER. South Africa, 2n = 40 (de Wet, 1958). Nigeria, 2n = 36 J. H. Davies B53/63/OP!. W. Cameroun, 2n = c. 36 Brunt 1182!. Uganda, 2n = c. 36 Bowden!. Kenya, 2n = 36 Bogdan 5744!. Madagascar, 2n = 36 (Tateoka, 1965c)!. Kenya, 2n = 30 (Celarier, 1956). Madagascar, 2n = 30 (Celarier, 1956)!. Brazil, 2n = 30 (Celarier & Harlan, 1957); n = 20 (Mehra, 1955).

HABITAT. Typically a tall savanna grass 2–2·5 m. high, often preferring the moister sites, but may be only 30–50 cm. high by roadside ditches and other frequented places. If grazed when young it will form a thick close sward.

DISTRIBUTION. A common species throughout tropical Africa. Less common in tropical America, and probably introduced elsewhere.

AUSTRALIA. Queensland: South Kennedy distr. (June), *Rounsell* 3; Gladstone (Mar.), *Blake* 12801; Maryborough (Oct.), *Clemens*.

HAWAII. Hawaii Nat. Park, Keahou (Feb.), *Fagerlund & Mitchell* 390.

MAURETANIA. L. Rkiz (Sept.), *Sadio* 116.

SENEGAL. Cape Verde (Dec.), *Adam* 1841; Diénoundiella (Dec.), *Berhaut* 2713.

GAMBIA. Wallikunda (Sept.), *Macluskie* 23.

MALI. Bamako (Dec.), *Adam* 11309, 11409; Koulikoro (Sept.), *Rogeon* 228; Macina (Dec.), *Duong* 1509; Kadigue to Banguita, *Davey* 8.

UPPER VOLTA. Leo to Po (Oct.), *Rose Innes* GC 31469.

PORTUGUESE GUINEA. Canquelifá (Dec.), *Pereira* 3617.

GUINEA. Seriba (Oct.), *Adam* 12710; La Kolenté (Nov.), *Chillou* 1030; Baffing valley (Nov.), *Pobéguin* 1816; Nzérékoré (June), *Adam* 5306; Timbo to Kankan (Dec.), *Adam* 2941.

SIERRA LEONE. Musaia (Feb., Mar.), *Deighton* 4205, 5448; Benekoro (Nov.), *Glanville* 318; Dumbaia (Jan.), *Morton & Gledhill* SL 573; Sefadu (Dec.), *Deighton* 3573; Senehun (Apr.), *Morton & Gledhill* SL 1963.

IVORY COAST. Nambonkaha (Nov.), *Leeuwenberg* 2048; Ferkessedougou (Nov.), *Leeuwenberg* 2018; Zuénoula (Oct.), *Adjanohoun* 309a.

GHANA. Lawra to Wa (Oct.), *Rose Innes* GC 32353; Navrongo to Wiaga (Oct.), *Ankrah* GC 20480; Pong Tamale (Dec.), *Morton* GC 9834; Elmina (Nov.), *Rose Innes* GC 31172; Accra, Achimota (Dec.), *Ankrah* GC 20115.

TOGO. Lomé, *Stage* 19.

DAHOMEY. Gouka (Mar.), *Froment* 1186; Kpinnou (Jan.), *Froment* 1133; Boukombe (Nov.), *Risopoulos* 1258; Cotonou to Allada (Oct.), *Risopoulos* 1237.

NIGERIA. Sokoto (Oct.), *Dalziel* 487; Gajibo (Dec.), *Davey* FHI 27123; Zaria (Nov.), *Thatcher* S. 523; Minna (Dec.), *Keay* FHI 37303; Gurum (Nov.), *Hepper* 1310; Ibadan (Nov.), *Keay* FHI 28138; Obudu (Dec.), *Tuley* 60.

W. CAMEROUN. Nkambe, Lus (Feb.), *Hepper* 1879; Wum (Nov.), *Brunt* 870; Bambui to Bamenda (June), *Brunt* 1182; Nyen (May), *Brunt* 1129; Victoria, *Maitland* 150a.

E. CAMEROUN. Maroua (Dec.), *Vaillant* 2473; Garoua (Oct.), *Vaillant* 2477; Sanguéré (Dec.), *Raynal* 12545; Bertoua (Dec.), *Baldwin* 13944; Yaounde, *Zenker & Staudt* 461; Douala to Loum (June), *Brunt* 1175.

CHAD. Ft. Archambault to Ft. Crampel (Nov.), *Chevalier* 10406; Bousso to Ft. Archambault (Nov.), *Chevalier* 10488.

CENTRAL AFRICAN REPUBLIC. Ndellé (Feb., Dec.) *Chevalier* 6829, 7671; Bouar (Dec.), *Boudet* 1550; Bambari (Nov.), *Tisserant* 2087; Bossembélé (Sept.), *Koechlin* 6104.

REPUBLIC OF CONGO. Without locality, *Burton, C. Smith*; Niari, *Thollon* 1078; Loudima (May), *Blauchon* 16; Brazzaville (July), *Chevalier* 4043.

CONGO REPUBLIC. Parc Nat. Garamba (Sept.), *Noirfalise* 755; Bagbele (Aug.), *Troupin* 2007; Kigenge (May), *Léonard* 4119; Kiwembe (May), *Cullens* 3618; Makayabo (Apr.), *Vanderyst* 3773; Elisabethville (May), *Symoens* 11011.

RWANDA. Mutara (May), *Troupin* 7235; Parc Nat. Kagera (Jan., Feb.), *Lebrun* 9578, *Troupin* 6232; Astrida (Apr.), *Van der Ben* 1534.

BURUNDI. Kirambo to Bweru (Mar.), *Van der Ben* 2485; Karuzi to Bweru (May), *Van der Ben* 2525.

SÃO THOMÉ. S. Antonio (July), *Moller* 152.

ANNOBON. Without locality, *Rose* 628; NW. corner (July), *F. Melville* 118.

SUDAN. Jebel Marra, Nyertete (Jan.), *Wickens* 1108; Rumbek (Feb.), *M. N. Harrison* 697; Kagelu (Mar.), *Myers* 6349; Lado (Mar.), *Speke & Grant*; Torit (Oct.), *J. K. Jackson* 874.

ERITREA. Soyra Mts. (Aug.), *Pappi* 1242.

ETHIOPIA. Without locality, *Schimper* 928; Adoa (Nov.), *Schimper* 1118; Gondar (Oct.), *Chiovenda* 2386; Bahar-Dar (Nov.), *Pichi-Sermolli* 77; Jimma (Oct.), *Mooney* 5956; Galla Sidama, *Piovano* 79.

UGANDA. West Nile, Pakwada (Nov.), *Langdale-Brown* 2370; Napenyenya (Sept.), *Dyson-Hudson* 110; Teso, Komolo Swamp (Mar.), *Michelmore* 1218; Entebbe (July), *Maitland* 15; Toro, Bwera (Feb.), *A. S. Thomas* 2761; Bushenyi (Jan.), *Snowden* 1283.

KENYA. Kitale (Sept.), *Bogdan* 1952; Kavirondo (Apr.), *Linton* 94; Meru (July), *Bogdan* 3750; Nairobi, *Dawson* 193; Malindi (Aug.), *Bogdan* 2564.

TANZANIA. Musoma, Nyamakachawe plains (Mar.), *Greenway, Turner & Allen* 10507; Kilimanjaro, *H. H. Johnston*; Uluguru Mts. (Apr.), *Schlieben* 3815; Milepa (Apr.), *Michelmore* 1163; Songea (Apr.), *Milne-Redhead & Taylor* 9843. Zanzibar: Kokotoni (Oct.), *Hildebrandt* 1075; Chaani (Dec.), *Greenway* 2614; Massazine (June), *Faulkner* 2606. Pemba I. (Oct.), *Burtt-Davy* 2510/29.

MOZAMBIQUE. Angónia (May), *Mendonça* 4178, 4188; Massangulo (Apr.), *Sousa* 1411; Nhamarroi (Sept.), *Torre* 3499; Mouth of Kongoni R. (Feb.), *Kirk*.

MALAWI. Karonga (June), *G. Jackson* 551; Lilongwe (Apr.), *G. Jackson* 471; Salima (Apr.), *G. Jackson* 750; Zomba, *Cormack* 475; Cholo (Dec.), *Wiehe* N371.

ZAMBIA. Mwengo (June), *Bullock* 3949; Mwinilunga, Matonchi farm (Oct.), *Milne-Redhead* 2603; Mufulira (May), *Eyles* 8379; Mongu (June), *Verboom* 1394; Choma, Muzoka (May), *Astle* 2486.

RHODESIA. Miami (Mar.), *Wild* 1767; Wankie (Mar.), *Brain* 3278; Salisbury (Feb.), *Eyles* 1519; Matopos (Feb.), *Rattray* 1596; Melsetter (Apr.), *Armitage* 84/55.

ANGOLA. Congo Yala (Mar.), *Gossweiler* 8898; Loanda, *Gossweiler* 402; Mutenda (Aug.), *Gossweiler* 8337; Calemba Is., *Welwitsch* 7409; Princeza Amelia, *Gossweiler* 3927.

MADAGASCAR. Majunga (May), *Afzelius*; Tamatave to Antananarivo (May), *Meller*; Tananarive (June), *Tateoka* 3523; Antsirabe (July), *Humbert & Swingle* 4648; Zombitsy, *Humbert* 29669; Ambongo, *Perrier de la Bâthie* 177.

COMORO Is. *Humblot* 148, *Bosser* 18023, 18317.

MAURITIUS. Without locality, *Ayres, Blackburn, Bouton, Vaughan* A16.

SEYCHELLES. Mahé Is., *Boivin* ; Victoria (Nov.), *Dupont* 2; Niole (Nov.), *Jeffrey* 366; Prashin Is. (Oct.), *Vesey-FitzGerald* 5630; Curicuse Is. (Jan.), *Vesey-FitzGerald* 5438.

SOUTH AFRICA. Transvaal: Letaba (Apr.), *Scheepers* 215.

U.S.A. Florida, Okeechobee (Dec.), *Moldenke* 243.

MEXICO. Veracruz, Campo Cotaxtla (Nov.), *McKee* 10980.

EL SALVADOR. Morazán, Divisadero (Dec.), *Tucker* 554; San Miguel, Hacienda Potrero Santo (Feb.), *Tucker* 953.

HONDURAS. Agua Azul (Apr.), *C. V. Morton* 7696.

NICARAGUA. Santa Maria de Ostuma (Dec.), *Hawkes* 2193.

COSTA RICA. El General (Jan.), *Skutch* 4035; Villa Colon (Feb.), *Lathrop* 5548.

ST. KITTS. Mount Pleasant (Jan.), *Box* 131.

GUYANA. Ebini (Jan., Apr.), *S. G. Harrison* 1775, *Harrison & Persaud* 701.

FRENCH GUIANA. Cayenne (Jan.), *Hoock* 1074.

BRAZIL. Belem (Aug.), *Glaziou* 9054; Loreto (June), *Eiten & Eiten* 4871; Paraguaçu Paulista (Feb.), *Clayton* 4564; Sertãozinho (Jan.), *Clayton* 4134; Matão (Jan.), *Clayton* 4118.

PANAMA. Chepo (Jan.), *Hunter & Allen* 80.

COLOMBIA. Loretoyacu R. (Oct.), *Schultes & Black* 8489.

6b. var. **siamensis** *W. D. Clayton*, var. nov.; a varietate typica pari race-morum 6–7-aristato, basi longiore racemorum 3–5 mm. longa, paribus spicularum homogamarum 1–2 instructa, et callo longiore distinguenda. Typus: Thailand, *Smitinand & Abbe* 6188 (K, holotype).

Racemes 6–7-awned per pair; upper raceme-base 3–5 mm. long, glabrous, bearing 1–2 homogamous pairs. *Sessile spikelets* 4–5 mm. long; callus slender, narrowly truncate, 1–1·2 mm. long. Otherwise as in var. *rufa*.

HABITAT. Roadsides and clearings in deciduous forest.

DISTRIBUTION. Thailand and adjacent countries.

BURMA. Mohnyin (Nov.), *U Thein Lwin* 618; Maymyo Plateau (Oct.), *Lace* 4291, 4382.

THAILAND. Chiengmai, Mê Teng (Nov.), *Kerr* 6672; Doi Suthep (Nov., Dec.), *Kerr* 2253, *Smitinand* 3915, 4005, *Smitinand & Abbe* 6188, *Sorensen, Larsen & Hansen* 2569; Lomkao (May), *Smitinand* 2614; Surin, Sangka (Jan.), *Kerr* 8932.

LAOS. Muang Awm (Apr.), *Kerr* 20972.

The greater part of this very common species is not difficult to recognize, being well characterized by the glossy lower glume of the sessile spikelet with its indumentum of sparse, rather stiff, rufous bristles; by its short obtuse callus; and by its dense racemes of 9–14 awns per pair.

It is however most variable in form, as perhaps can be expected from the long synonymy and the variety of recorded chromosome numbers. Some of the variants have been listed under the heading 'minor variations', where the possibility of sub-dividing the species was rejected as impracticable, at least in the present state of our knowledge. It is true that some botanists, with great experience of their region, have learned to distinguish several forms in the field, but I doubt whether these taxa amount to more than local ecotypes.

Although the central core must apparently remain indivisible, variation at the periphery is of a different order and requires that some attempt be made to limit the circumscription of the species. The nature of the problem is illus-trated in figure 18, where four of the differential characters are plotted for *H. poecilotricha*, *H. dichroa*, and an equal-sized sample of *H. rufa* taken from the Congo. The presence of three populations is undeniable but, translated into taxonomic terms, it is impossible to specify the precise point at which we

should cease to use one name and commence to use another. Despite the inde-
terminacy of the boundary, I believe that the taxonomic distinction is valid
and should be upheld.

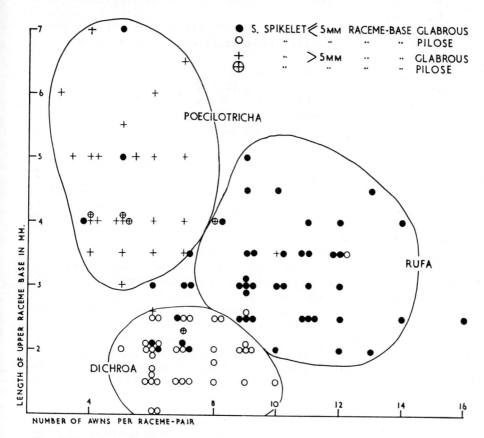

FIG. 18. *Hyparrhenia poecilotricha, H. dichroa* and a sample of *H. rufa* from the Congo. Length of
upper raceme-base plotted against number of awns per raceme-pair.

A further difficulty arises in Thailand. The typical African form of *H. rufa*
does not seem to occur there; instead there is found a remarkably homogeneous cluster of specimens lying on the boundary between *H. rufa* and
H. poecilotricha. Similar specimens have been collected from several widely
dispersed localities in Africa, and have there been regarded as peripheral forms
of *H. rufa*, for they merge by imperceptible stages with the main body of that
species. It is as though only a marginal fragment from the gene pool of *H. rufa*
has established itself in distant lands. A coherent population such as this would
seem to require taxonomic recognition (Stapf has annotated the Kew sheets
'*H. floribunda*'), but species rank is inappropriate for it cannot be adequately
separated from the African populations. Such a situation is not uncommon in
taxonomy, and has no solution which is both logical and taxonomically satisfactory. I have adopted the device of designating the Asiatic population as a
variety, without attempting to detach similar African specimens from the main
population of *H. rufa*.

7. **Hyparrhenia dichroa** *(Steud.) Stapf* in Prain, Fl. Trop. Afr. 9: 302 (1919).

Andropogon bicolor Nees, Fl. Afr. Austr. 1: 113 (1841), *non* Roxb. (1820). Type: South Africa, *Drège* (LUB, holotype destroyed during second world war; K, ? isotype).

A. dichroos Steud., Syn. Pl. Glum. 1: 389 (1854); Hack. in DC., Monogr. Phan. 6: 622 (1889); Stapf in Thistleton-Dyer, Fl. Cap. 7: 360 (1898). Based on *A. bicolor* Nees.

Sorghum bicolor (Nees) Kuntze, Rev. Gen. Pl. 2: 790 (1891), *non* Moench (1794).

Cymbopogon luembensis De Wild. in Bull. Jard. Bot. Brux. 6: 14 (1919). Type: Congo Republic, *Hock* (BR, holotype; K, isotype).

Andropogon luembensis De Wild., *l. c.* (1919), *in synon.*

Hyparrhenia luembensis (De Wild.) Robyns, Fl. Agrost. Congo Belge 1: 183 (1929).

H. fastigiata Robyns, *l. c.*: 164 (1929), & in Bull. Jard. Bot. Brux. 8: 231 (1930). Type: Congo Republic, *Scaetta 7* (BR, holotype).

Perennial; culms tufted, stout, up to 3 m. high. *Leaf-sheaths* glabrous; ligule 1 mm. long; blades rigid, up to 60 cm. long and 8 mm. wide. *Spathate panicle* copiously branched, 20–60 cm. long; spatheoles narrowly lanceolate, 2–3·5 cm. long and 1·5–2·5 mm. wide in profile, at length reddish; peduncles mostly 1–2·5 cm. long, half to almost as long as the spatheole, glabrous or white pilose towards the top. *Racemes* not deflexed (or very rarely so), 1–1·5(–2) cm. long, 6–9-awned per pair, glabrescent to fulvous, terminally exserted; raceme-bases unequal, the lower very short, the upper 1·5–2·5 mm. long, hirtellous or sometimes glabrous, oblique at the tip. *Homogamous pairs* 1 at the base of the lower or both racemes; spikelets resembling the pedicelled. *Sessile spikelets* lanceolate, 4–5 mm. long; callus 0·4–0·8 mm. long, cuneate, obtuse; lower glume yellowish brown or violet-tinged, usually glossy, the hairs fulvous but often very pallid and scanty; awns 2–3 cm. long, the column rufously pubescent. *Pedicelled spikelets* narrowly lanceolate, 3–5 mm. long, acute or mucronate at the apex; pedicel tooth short, broadly triangular.

MINOR VARIATIONS. The upper raceme-base is usually sparsely hirtellous with soft hairs, but it may be glabrous, and is sometimes more densely pilose with stiffish bristles. In extreme cases it approaches the condition found in *H. gazensis*; indeed *Eyles* 8391 from Zambia resembles that species in all but the rufous hairs on the sessile spikelet. In a very few specimens the spikelet indumentum is white. The awn is occasionally shorter than usual, and in extreme examples (*Scaetta 7* and *28f* from the Congo Republic, and *A. S. Thomas* 681 and 1428 from Uganda) is only 9–10 mm. long.

HABITAT. A species of tall savanna grassland, especially the moister sites in *Brachystegia* woodland; also along roadsides and on old fallow land.

DISTRIBUTION. Southern Africa from Tanzania to Natal, and also in Sudan and Ethiopia (Map. 4, p. 61).

CONGO REPUBLIC. Busenene (Dec.), *van Ysacker* 130; Tshibinda, *Scaetta 7*, *28f*; Jadotville (Apr.), *Bredo* 2734; Luembe valley (Feb.), *Hock*.

RWANDA. Parc Nat. Kagera (June), *Troupin* 3589.

SUDAN. Jebel Marra (June), *Steele* 45; Wadi Azum (Jan.), *J. K. Jackson* 2525; Nyertela (May), *Blair* 2; Kulme (July), *Lynes* 625; Gallabat (Jan.), *J. K. Jackson* 3404.

ETHIOPIA. Tucur Dinghia (Jan.), *Pichi-Sermolli* 322; Zara Enda Michael (Mar.), *Pichi-Sermolli* 331.

UGANDA. Toro, Bwamba Pass (Sept., Nov.), *A. S. Thomas* 681, 1428.

KENYA. Uasin Gishu, *McDonald* 844.

TANZANIA. Mpwapwa (May), *van Rensberg* 446; Morogoro (Mar.). *Schlieben* 3657; Kampawanda (May), *Vesey-FitzGerald*; Ihosio (May), *Vesey-FitzeGrald*; Ruponda (July), *Semsei* in FH 2161. Zanzibar, *Sacleux* 411.

MOZAMBIQUE. Massingire (Aug.), *Torre* 4475; Vila Gouveia (Oct.), *Torre* 6093; Grudge to Vila Pery (June), *Torre* 2829; Chimoio (Nov.), *Pedro* 4576; Spungabera (June), *Torre* 4247.

MALAWI. Mzimba, *S. G. Wilson* 18*a*; Kota-Kota, Nchisi Mt. (Aug.), *Brass* 17090; Zomba (Aug.), *G. Jackson* 112; Shire highlands, *Buchanan* 39; Chikwawa (Oct.), *Lawrence* 73.

ZAMBIA. S. Mwinilungu distr. (Sept.), *Trapnell* 1611; Solwezi (June), *Milne-Redhead* 475; Mumbwa, *Macaulay*; Mpanza (Apr.), *E. A. Robinson* 185; Kalomo (June), *Mitchell* 7/55.

RHODESIA. Victoria Falls (May), *F. A. Rogers* 1327; Gokwe, Sessami (Apr.), *Bingham* 606; Hartley (May), *Rattray* 1465; Salisbury (Mar.), *Sturgeon* 57589; Chuhanje Range (June), *Wild* 3472.

SOUTH AFRICA. Transvaal: Punda Maria (Apr.), *Codd* 5963; Blouberg (Apr.), *Strey & Schlieben* 8597; Nelspruit (Oct.), *Shantz* 318. Natal: Without locality (Nov.), *Buchanan* 303; Inyezaan R. (Aprl), *J. M. Wood* 3927. Cape Province: St. John's R. to Umtsicaba R., *Drége* 4357.

In southern Africa, *H. rufa* merges with plants having almost identical spikelet characters, but with the panicle noticeably different in appearance. A more detailed analysis reveals that the racemes have fewer awns per pair, are borne on shorter peduncles, and tend to remain within the spatheole which is shorter and broader; moreover the upper raceme-base is often hirtellous. Together these characters constitute a marked departure from typical *H. rufa*, but separately they intergrade with the latter and the boundary is by no means clear (Fig. 18, p. 67). To what extent this population is genetically distinct remains, at present, a matter for conjecture. Meanwhile it is clearly unsatisfactory to lump it with *H. rufa*, and I have accordingly treated it as a separate species.

The hirtellous raceme-base of *H. dichroa* has sometimes led to confusion with *H. schimperi*, the range of spatheole and awn-length being much the same in both species. *H. schimperi* however has glabrous or white-haired spikelets, and a shorter, more densely pilose, raceme-base terminating in a little scarious lobe.

Specimens with very scanty or pallid spikelet hairs may be mistaken for *H. gazensis* or *H. finitima*. The former is a much less robust plant with a longer, stiffly pilose, raceme-base; the latter has a pungent callus.

Specimens with unusually short awns have been separated as a distinct species, *H. fastigiata*. Although there is some evidence for a disjunction in the frequency distribution of awn lengths, there are no convincing supporting characters, and I do not feel that a distinction at species level is justified.

8. **Hyparrhenia poecilotricha** (*Hack.*) *Stapf* in Prain, Fl. Trop. Afr. 9: 309 (1918).

Andropogon poecilotrichus Hack. in Bol. Soc. Brot. 3: 138 (1885), & in DC., Monogr. Phan. 6: 638 (1889). Type: Angola, *Newton* (K, isotype).

Sorghum poecilotrichum (Hack.) Kuntze, Rev. Gen. Pl. 2: 792 (1891).

Andropogon buchananii Stapf in Thisteleton-Dyer, Fl. Cap. 7: 362 (1898). Type: South Africa, *Buchanan* (K, holotype).

A. pleiarthron Stapf, *l. c.*: 364 (1898). Type: South Africa, *Nelson* 15 (K, holotype).

Cymbopogon pleiarthron (Stapf) Stapf ex Burtt-Davy in Ann. Transvaal Mus. 3: 121 (1912).

Hyparrhenia buchananii (Stapf) Stapf ex Stent in Bothalia 1: 249 (1924).

H. familiaris var. *pilosa* Robyns, Fl. Agrost. Congo Belge 1: 176 (1929), & in Bull. Jard. Bot. Brux. 8: 236 (1930). Type: Congo Republic, *Vanderyst* 6090 (BR, holotype).

Perennial; culms 60–130 cm. high. *Leaf-sheaths* glabrous, or very rarely loosely pilose; ligule 1·5 mm. long; blades rigid, up to 30 cm. long and 3 mm. wide. *Spathate panicle* lax and open, about 30 cm. long; spatheoles linear, 4–8 cm. long; peduncles about as long as the spatheole, with or without spreading white or yellowish hairs towards the top. *Racemes* not, or sometimes tardily, deflexed, 1·5–2 cm. long, 4–7-awned per pair, fulvous, terminally exserted; raceme-bases unequal, the upper 3·5–7 mm. long, glabrous or rarely with a few hairs, usually glabrous at the foot. *Homogamous pairs* 1 (very rarely 2) at the base of the lower raceme, and 2 (or sometimes only 1) at the base of the upper, the number usually variable within the same panicle; spikelets awnless, otherwise similar to the pedicelled. *Sessile spikelets* narrowly lanceolate, 5·5–7 mm. long, rarely less; callus acute to pungent, 1–2 mm. long; lower glume pubescent with fulvous or yellow hairs; awn 2·5–4 cm. long, the column pubescent with rufous hairs. *Pedicelled spikelets* narrowly lanceolate, 4–7 mm. long, usually with an awn-point up to 2 mm. long; pedicel tooth triangular, up to 0·2 mm. long.

MINOR VARIATIONS. The degree of hairiness of the racemes varies considerably, some specimens approaching *H. nyassae*, others resembling *H. rufa*. *Dyson-Hudson* 267 from Uganda is an extreme example with almost glabrous spikelets; its long upper raceme-base avoids confusion with *H. finitima*. The awn hairs are typically 0·5 mm. long, but may be longer in the specimens most resembling *H. filipendula*.

HABITAT. Savanna woodlands and dambo margins.

DISTRIBUTION. Eastern Africa from Sudan to Natal; only 5 specimens recorded from west of the Cameroun mountain range (Map 4, p. 61).

GUINEA. Labé (Nov.), *Adames* 410; Timbo (Dec.), *Adam* 2673; Nzérékoré to Beyla (May), *Adam* 5050.

GHANA. Gambaga (Oct.), *Rose Innes* GC 30764; Kintampo (Dec.), *Vigne* FH 3201.

CAMEROUN. Bambui (June), *Brunt* 1185; Métié (Apr.), *Raynal* 10861.

REPUBLIC OF CONGO. Loutété (June), *Blauchon* 53; Madingou (July), *Blauchon* 66.

CONGO REPUBLIC. Parc Nat. Garamba (Oct.), *de Saeger* 1472; Bogoro to Ngeti (Aug.), *H. B. Johnston* 1106; Walungu (Feb.), *Laurent* 631; Kampilikwe to Kasiki (June), *Symoens* 4670; Elisabethville (June), *Gathy* 732.

SUDAN. Jebel Marra (June, Oct.), *Blair* 65, 185; Torit, Farajok (Oct.), *J. K. Jackson* 362.

UGANDA. Karamoja, Morunyagai (July), *Dyson-Hudson* 267; Toro Game Reserve (Nov.), *Buechner* 6; Masaka (May), *Chandler* 1633; Sese Is., Mugoye (Feb., June), *A. S. Thomas* 56, 834.

KENYA. Thika (Jan., June), *Bogdan* 810, 1491, 3739; Fourteen Falls (Jan.), *Verdcourt* 2619; Nairobi (Jan.), *Bogdan* 1634; Gazi (Aug.), *Drummond & Hemsley* 3824.

TANZANIA. Bukoba, *Williams* 88; Kasembe (Apr.), *Pêtre*; Mbozi, *Jacobsen* 6.

MOZAMBIQUE. Angónia (May), *Mendonça* 4187; Vila Machado to Vila Pery (Sept.), *Torre* 5925.

ZAMBIA. Abercorn, Kawimbe (Mar.), *McCallum Webster* A235; Mporokoso (Apr.), *Phipps & Vesey-FitzGerald* 3142a; Luwingu (May), *Astle* 3025; Mazabuka (Jan.), *Astle* 2925; Livingstone (Feb.), *Mitchell* 24/84.

RHODESIA. Marandellas, Digglefold (June), *Corby* 123; Matopos (Feb.), *Rattray* 1608.

ANGOLA. Vila Luzo (May), *Gossweiler* 11277; Humpata (Mar.), *F. Newton*.

SOUTH AFRICA. Transvaal: Kranskop, *Nelson* 15. Natal: Umpumulo, *Buchanan*.

H. poecilotricha presents a difficult problem, for it has a rather wide, though continuous, range of variation which effectively bridges the gap between four other species.

At one end of its range of variation it tends to merge with *H. rufa* (Fig. 18, p. 67) and *H. nyassae*, and there is no single reliable discriminatory character. On the other hand the long upper raceme-base, relatively few spikelets per raceme-pair, fairly long sessile spikelet with a long narrow callus, glabrous basal sheaths, 1–2 homogamous pairs at the base of the upper raceme, and aristulate pedicelled spikelets provide a suite of differential characters, which can scarcely be ignored. I have therefore accepted the separation of this species, when in doubt giving most weight to the lengths of the sessile spikelet and the upper raceme-base.

At the other extreme there are specimens which might easily be included in *H. filipendula* or *H. familiaris*, were it not for their rufous indumentum; the type of *H. familiaris* var. *pilosa*, with distinctly tawny hairs, is a good example. As a rule this character gives little trouble, but there is sometimes difficulty in separating *H. poecilotricha* from a few specimens of *H. filipendula* var. *pilose* and *H. quarrei* in which the hairs have acquired a yellowish tinge; such specimens are commonly mixed on the same sheet with panicles having pure white spikelet hairs. This is probably a matter of introgression, in which case a clear-cut separation is not to be expected.

It is difficult to avoid the suspicion that *H. poecilotricha* is made up of a hodge-podge of hybrids between members of the *Rufa* and *Filipendula* groups. Its rarity in West Africa, where *H. filipendula* is also uncommon, lends support to this hypothesis. One is tempted to essay a subdivision according to the supposed parentage of its components, but the characters available are insubstantial, and I believe that the present broad treatment is preferable to a spuriously precise fragmentation into vaguely circumscribed micro-taxa.

9. **Hyparrhenia gazensis** (*Rendle*) *Stapf* in Prain, Fl. Trop. Afr. 9: 301 (1918).

Cymbopogon gazensis Rendle in Journ. Linn. Soc., Bot. 40: 226 (1911). Type: Rhodesia, *Swynnerton* 1637 (BM, holotype; K, isotype).

Andropogon gazensis (Rendle) Eyles in Trans. Roy. Soc. S. Afr. 5: 294 (1916).

Hyparrhenia snowdenii C. E. Hubbard in Bull. Misc. Inf. Kew. 1928: 38 (1928). Type: Uganda, *Snowden* 1150 (K, holotype).

Loosely tufted perennial; culms slender, 50–180 cm. high and up to 2 mm. in diameter, weakly erect or geniculately ascending, sometimes sprawling. *Leaf-sheaths* glabrous, or pubescent above; ligule up to 1 mm. long; blades 8–20 cm. long and 2–5 mm. wide, long acuminate, glabrous or thinly pilose. *Spathate panicle* narrow, loose, 10–35 cm. long and up to 5 cm. wide; spatheoles linear-lanceolate, 3–4 cm. long, glabrous, reddish-brown; peduncles 1–3·5 cm. long, $\frac{1}{2}$–$\frac{3}{4}$ the length of the spatheole, pilose with white hairs above. *Racemes* not deflexed, 4–5-awned per pair, 1–1·5 cm. long, laterally exserted; raceme-bases unequal, divergent, the upper filiform, (2–)2·5–3·5 mm. long, stiffly pilose with white hairs, subtruncate at the tip. *Homogamous pairs* 1 at the base of the lower or both racemes; spikelets narrowly lanceolate, 5 mm. long. *Sessile spikelets* linear-oblong to lanceolate-oblong, 4·5–5 mm. long; callus cuneate, 0·8–1·5 mm. long, acute; lower glume hispidulous; awn 2–3 cm. long, the column pubescent with fulvous hairs. *Pedicelled spikelets* linear-lanceolate, 5–6 mm. long, glabrous, terminating in a short mucro 1–2 mm. long; pedicel tooth short, broadly triangular.

CHROMOSOME NUMBER. South Africa, 2n = 30 (de Wet, 1958).

HABITAT. A ruderal on poor soils, following cultivation and along roadsides.

DISTRIBUTION. Centred upon Uganda and Transvaal, but extending into adjacent territories (Map 5).

CONGO REPUBLIC. Mahagi (Oct.), *H. B. Johnston* 1176; Katanga, Matimbala (May), *Bovone* 50.

UGANDA. Luwero to Wakyato (Apr.), *Langdale-Brown* 2061; Kepeka (Oct.), *H. B. Johnston* 1305; Budama, Samia (July), *Snowden* 1150; Busogo, Nawailogo swamp (Aug.), *Langdale-Brown* 2319.

KENYA. Lambwe valley, Kaniamwia escarpment (Aug.), *Makin* 147.

RHODESIA. Chirinda (June), *Swynnerton* 1637; Melsetter (Jan.), *Crook* 501.

SOUTH AFRICA. Transvaal: Louis Trichardt, *F. A. Rogers* 21124; Duiwelskloof (May), *Scheepers* 318; Tzaneen (Sept.), *Schlieben* 7201; *de Winter & Codd* 145; Pilgrim's Rest (Nov.), *Killick & Strey* 2455. Natal: Hlabisa (Apr.), *C. J. Ward* 2548.

H. gazensis closely resembles *H. dichroa* in general appearance, but is distinguished by its weak culms, fewer awns per raceme-pair, almost glabrous spikelets, and longer raceme-base clothed in stiff hairs. The species has many affinities with sect. *Pogonopodia* but, although the upper raceme-base is stiffly pilose, it is also long and filiform so that the species cannot be satisfactorily accommodated in that section.

10. **Hyparrhenia finitima** (*Hochst.*) *Anderss. ex Stapf* in Prain, Fl. Trop. Afr. 9: 299 (1918).

Andropogon finitimus Hochst. in sched., Schimp. Iter Abyss. 2: 1797 (1844); A. Rich., Tent. Fl. Abyss. 2: 465 (1851); Hack. in DC., Monogr. Phan. 6: 637 (1889). Type: Ethiopia, *Schimper* 1797 (K, isotype).

Cymbopogon finitimus (Hochst.) T. Thoms. in Speke, Journ. Discov. Source Nile: 652 (1863).

Sorghum finitimum (Hochst.) Kuntze, Rev. Gen. Pl. 2: 791 (1891).

Hyparrhenia rhodesica Stent & Rattray in Proc. Rhod. Sci. Ass. 32: 14 (1933). Type: Rhodesia, *Perrotet* in *Eyles* 3069 (K, isotype).

H. hirta var. *garambensis* Troupin, Fl. Garamba 1: 47 (1956). Type: Congo Republic, *Troupin* 175 (BR, holotype).

Tufted perennial; culms 1–2 m. high. *Leaf-sheaths* at base of plant pubescent to hirsute along the margins and sometimes also on the back; ligule up to 2 mm. long; blades up to 60 cm. long and 8 mm. wide. *Spathate panicle* up to 60 cm. long, copiously branched and usually rather dense; spatheoles narrowly lanceolate, 2·5–4 cm. long, embracing the racemes; peduncles $\frac{1}{3}$–$\frac{2}{3}$ as long as the spatheole, pilose with white hairs above. *Racemes* not deflexed, 2–6-awned per pair, 1–1·5 cm. long, yellowish, laterally exserted; raceme-bases unequal, the upper filiform, 1·5–2·5 mm. long, hirtellous to pilose with white hairs, glabrous at the foot. *Homogamous pairs* 1 at the base of the lower or both racemes; spikelets narrowly lanceolate, 6–8 mm. long. *Sessile spikelets* narrowly

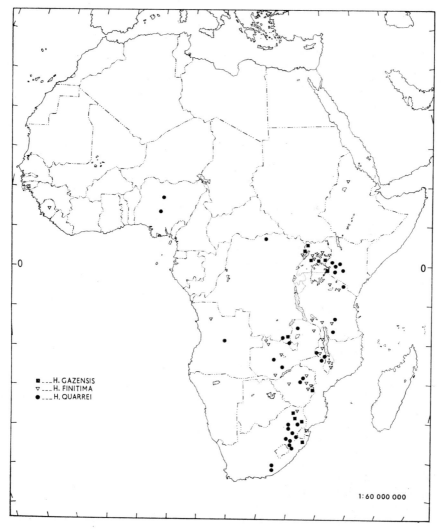

MAP 5. Distribution of *Hyparrhenia gazensis*, *H. finitima* and *H. quarrei*.

lanceolate, 5·5–6 mm. long; callus linear, 1–2 mm. long, acute to very narrowly truncate at the tip; lower glume yellowish, glabrous to hispidulous with white hairs; awn 2·5–4 cm. long, the column puberulous with fulvous hairs. *Pedicelled spikelets* narrowly lanceolate, 5–6 mm. long, glabrous or rarely hispidulous, terminating in an awn 2–5 mm. long; pedicel tooth triangular to subulate, up to 0·2 mm. long.

MINOR VARIATIONS. Specimens from Uganda have glabrous basal sheaths, a shorter callus (0·8–1 mm. long), and merely an awn-point about 1 mm. long terminating the pedicelled spikelets. These characters are not sufficiently distinctive to justify separation, and the population is best regarded as forming one end of the range of variation of *H. finitima*. The type of *H. rhodesica* also has awnless pedicelled spikelets, but in addition the short (1·5 mm.) raceme-bases are partially fused.

HABITAT. A ruderal species of roadsides, waste land and farm fallow.

DISTRIBUTION. Rather widely distributed, but mainly in south tropical Africa, with a lesser centre in Uganda. Note the general similarity with the distribution of *H. gazensis*. (Map. 5, p. 73).

SIERRA LEONE. Mateboi (Nov.), *Jordan* 843.
CONGO REPUBLIC. Parc. Nat. Garamba (Feb.), *Troupin* 175, 1695, 1904.
ETHIOPIA. Djeladjeranne (Oct.), *Schimper* 1797; Mai Zabri to Mai Berentia (Dec.), *Chiovenda* 3221.
UGANDA. Masindi (July), *Langdale-Brown* 1401, 1413; Kiwoko to Wakyato (May), *Langdale-Brown* 2090; Kepeka (Oct.), *H. B. Johnston* 1306; Sabei (Dec.), *A. S. Thomas* 2615.
TANZANIA. Serengeti, Seronera Dam (Apr.), *Greenway* 10072; L. Manyara Nat. Park (Mar.), *Greenway & Kanuri* 11396; Mbosi, *R. M. Davies* 520; Mbeya, Lupa N.F.R. (Apr.), *Boaler* 541; Rukisa, Ilemba Gap (Mar.), *McCallum Webster* T75.
MOZAMBIQUE. Namaacha hills (Sept.), *Hornby* 1052.
MALAWI. Mzimba (Apr.), *G. Jackson* 1610; Kotakota (Oct.), *G. Jackson* 122; Lilongwe (Apr.), *G. Jackson* 460, 469; Zomba, *Cormack* 3; Mlanje (June), *J. D. Chapman* 649.
ZAMBIA. Chinsali (May), *Astle* 3054; Mwinilunga (Nov.), *Duff* 1148; Mwinilunga district, Matonchi farm (Nov.), *Milne-Redhead* 3198; Chisamba to Broken Hill (June), *Trapnell* 2034; Choma, Mochipapa (Apr.), *Astle* 2990.
RHODESIA. Gokwe (Mar.), *Bingham* 487; Salisbury (Mar.), *Sturgeon* in GH 57586; Marandellas (May), *Rattray* in GH 20590; Chiduku Reserve (Apr.), *R. M. Davies* 1083; Umtali (May), *Perrott* in *Eyles* 3069; Sabi valley, Rupisi (Oct.), *D. A. Robinson* 415.
ANGOLA. Caghug, *Welwitsch* 2838.
SOUTH AFRICA. Transvaal: Louis Trichardt (Mar.), *van Vuuren* 1685.

Closely allied to *H. gazensis*, from which it is distinguished by the robust culms, slender callus, awned pedicelled spikelets, and shorter raceme-base; the Uganda specimens are in some respects intermediate between the two species. It also bears some resemblance to *H. hirta* and *H. filipendula*, being separated from the former by its glabrous spikelets, few-awned raceme-pairs and copious panicles; and from the latter by its short raceme-base and single homogamous pair.

11. **Hyparrhenia wombaliensis** (*Vanderyst ex Robyns*) *W. D. Clayton*, comb. nov.

Andropogon wombaliensis Vanderyst ex Robyns, Fl. Agrost. Congo Belge 1: 128 (1929); Vanderyst in Bull. Soc. Bot. Belge 55: 39 (1923), *nom. prov.* Type: Congo Republic, *Vanderyst* 4251 (BR, holotype; K, isotype).

A. wombaliensis var. *ciliatus* Robyns, Fl. Agrost. Congo Belge 1: 128 (1929). Type: Congo Republic, *Vanderyst* 15142 (BR, holotype).

Densely caespitose perennial; culms slender, 40–60 cm. high. *Leaf-sheaths* glabrous or the lower sparsely hirsute; ligule 0·5–1 mm. long; blades very narrow, up to 25 cm. long and 1·5 mm. wide, often convolute and filiform, glabrous or loosely hirsute beneath. *Inflorescence* a single pair of racemes borne at the summit of the culm, sometimes with a second pair on a long slender branch from the penultimate node; spatheoles very slender, 4–10 cm. long, tightly rolled about the peduncle; peduncle about twice as long as the spatheole, glabrous. *Racemes* not deflexed, 2–3 cm. long, 9–12-awned per pair, loose, pale green to violet; internodes 3·5–5 mm. long; raceme-bases unequal, the upper 5–8 mm. long, glabrous. *Homogamous pairs* 1 at the base of the lower raceme only, resembling the pedicelled spikelets. *Sessile spikelets* narrowly lanceolate, 5–6 mm. long, glabrous or sparsely hispidulous; callus acute, 1 mm. long; upper lemma lobes reduced to obscure shoulders; awn 8–15 mm. long, the column pubescent with shaggy hairs 0·2–0·3 mm. long. *Pedicelled spikelets* narrowly lanceolate, 5–6 mm. long, glabrous, acute; pedicel tooth very short, subulate.

HABITAT. A species of waterlogged sandy soils.

DISTRIBUTION. Confined to the lower reaches of the Congo basin (Map 6, p. 76).

CONGO REPUBLIC. Stanley Pool, Wombali, *Vanderyst* 4251; Kisantu, Nto-Mbombo (Dec.), *Callens* 3893; Luvu (Dec.), *Callens* 3805; Kutu (Oct.), *Flamigni*; Popokabaka (Sept.), *Vanderyst* 15142.

The species resembles *H. hirta*, although the spikelets are glabrous. It may be readily recognized by the very scanty inflorescence, which is commonly reduced to a single pair of racemes. In this respect it resembles many species of *Andropogon*, but the sessile spikelet callus is quite clearly applied obliquely to the internode tip, nor does the lower glume show any trace of lateral keels. The species has accordingly been transferred to *Hyparrhenia*. A unique feature, found in no other species of *Hyparrhenia*, is the almost complete reduction of the lateral lobes of the upper lemma.

12. **Hyparrhenia hirta** (*L.*) *Stapf* in Prain, Fl. Trop. Afr. 9: 315 (1918).

Andropogon hirtus L., Sp. Pl. ed 1: 1046 (1753); Hack. in DC., Monogr. Phan. 6: 618 (1889); Stapf in Thistelton-Dyer, Fl. Cap. 7: 335 (1898). Type: Italy, *Burser* I. 119 (UPS, holotype; K, microfiche).

A. pilosus Dufour in Roem. & Schult., Sp. Pl. 2: 819 (1817), *non* Klein ex Willd. (1806). Type: Italy, *Dufour*.

A. pubescens Vis. in Flora 12 (1) Erg.: 3 (1829), *non* Dryand. (1789). Type: Yugoslavia, *Visiani* (FI, holotype).

Trachypogon hirtus (L.) Nees, Agrost. Bras.: 346 (1829).

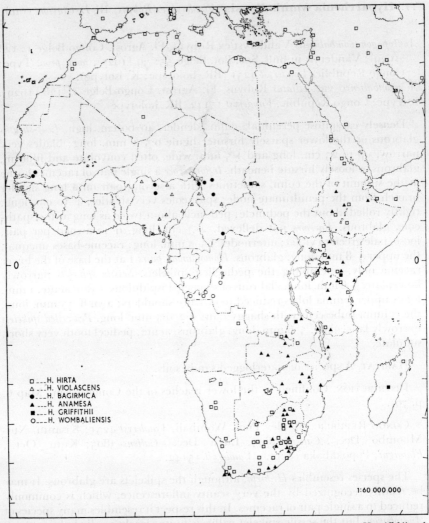

MAP. 6. Distribution of *Hyparrhenia hirta, H. violascens, H. bagirmica, H. anamesa, H. griffithii* and *H. wombaliensis.*

Andropogon giganteus Ten., Fl. Nap. 5: 285 (1836) Type: Italy, *Dufour.*

A. sinaicus Del. in Ann. Sci. Nat. sér. 2, 7: 285 (1837). Type: grown at Montpellier from seed from Mt. Sinai.

A. hirtus var. *glaucus* Schrad. in Linnaea 12: 475 (1838). Based on *A. pubescens* Vis.

A. podotrichus Hochst. in sched., Schimp. Iter Abyss. 2: 1056 (1842). Type: Ethiopia, *Schimper* 1056 (K, isotype).

A. hirtus var. *longearistatus* Willk. & Lange, Prodr. Fl. Hisp. 1: 47 (1861). Based on *A. pubescens* Vis.

Cymbopogon hirtus (L.) T. Thoms. in Speke, Journ. Discov. Source Nile: 652 (1863).

Hyparrhenia podotricha (Hochst.) Anderss. in Schweinf., Beitr. Fl. Aethiop.: 310 (1867), *nom. inval.* (genus not yet described).

Heteropogon hirtus (Hochst.) Anderss. *l. c.* 300 (1867).

H. pubescens Anderss. *l. c.* (1867). Based on *Andropogon pubescens* Vis.

Andropogon hirtus var. *pubescens* (Anderss.) Vis., Fl. Dalm. Suppl.: 150 (1872).

A. ambiguus Gennari ex Barbey, Fl. Sard. Comp.: 190 (1884), *non* Michx. (1803). Type: Sardinia, *Gennari.*

A. hirtus var. *podotrichus* (Hochst.) Hack. in DC., Monogr. Phan. 6: 620 (1889).

Sorghum hirtum (L.) Kuntze, Rev. Gen. Pl. 2: 792 (1891).

Andropogon transvaalensis Stapf in Thistleton-Dyer, Fl. Cap. 7: 363 (1898). Type: South Africa, *E.S.C.A. Herb.* 301 (K, holotype; WAG, isotype).

A. collinus Lojac., Fl. Sic. 3: 272 (1909). Type: Sicily, *Tineo.*

Cymbopogon pubescens (Anderss.) Fritsch, Exkursionsfl. Oesterr. ed. 2: 47 (1909).

C. transvaalensis (Stapf) Stapf ex Burtt-Davy in Ann. Transvaal Mus. 3: 122 (1912).

C. modicus De Wild. in Bull. Jard. Bot. Brux. 6: 16 (1919). Type: Congo Republic, *Bequaert* 310 (BR, holotype).

Andropogon modicus De Wild., *l. c., in synon.*

Hyparrhenia pubescens (Anderss.) Chiov., Pl. Nov. Aethiop.: 20 (1928).

H. modica (De Wild.) Robyns, Fl. Agrost. Congo Belge 1: 172 (1929).

Andropogon hirtus var. *glaucus* Nábělek in Publ. Fac. Sci. Univ. Masaryk 111: 5 (1929), *non* Schrad. (1838). Type: Palestine, *Nábělek* 3132.

A. hirtus var. *glabriglumis* Oppenh., Fl. Transiord.: 145 (1931). Type: Jordan, *Aaronsohn* 254.

H. hirta var. *longearistata* (Willk. & Lange) Rothm. & Silva in Agron. Lusit. 1: 240 (1939).

H. hirta var. *podotricha* (Hochst.) Pichi-Sermolli in Miss. Lago Tana 7, Ric. Bot. 1: 174 (1951).

H. hirta f. *podotricha* (Hochst.) Maire & Weiller, Fl. Afr. Nord 1: 291 (1952).

H. hirta f. *pubescens* (Anderss.) Maire & Weiller, *l. c.* (1952).

H. hirta subsp. *pubescens* (Anderss.) Paunero in Anal. Bot. Cavan. 15: 430 (1957).

H. hirta f. *brachyphylla* Paunero *l. c.* (1957), *nom. nud.*, sine descr. lat.

H. hirta var. *villosa* Paunero *l. c.* (1957), *nom. nud.*, sine descr. lat.

H. hirta subsp. *villosa* Pign. in Arch. Bot. Biog. Ital. 34: 3 (1958). Type: Italy, *Pignatti.*

H. hirta var. *pubescens* (Anderss.) Rawi in Dept. Agr. Iraq Tech. Bull. 14: 209 (1964).

Caespitose perennial arising from short underground rhizomes; flowering culms wiry, typically 30–60 cm. high (1 m. or more in exceptionally robust plants), standing over a dense leafy tussock 10–20 cm. high. *Leaf-sheaths* compressed and keeled, glabrous or very rarely obscurely puberulous at the base of the plant; ligule up to 4 mm. long; blades narrowly linear to conduplicate and filiform, 2–15 cm. long and 1–2(–4) mm. wide, firm, flexuous, glaucous, glabrous or with some scattered hairs downwards, harshly scaberulous. *Spathate panicle* typically scanty, up to 30 cm. long, bearing 2–10 raceme-pairs or sometimes more; spatheoles linear-lanceolate, 3–8 cm. long, at length reddish; peduncles about as long as the spatheole, with or without spreading white hairs at the top. *Racemes* never deflexed, 2–4 cm. long, 8–13–(–16)-awned per pair, villous with white hairs, terminally exserted; raceme-bases unequal, the upper 2·5–5 mm. long, filiform, glabrous or more often pubescent to hirsute, with or without a white beard at the foot. *Homogamous pairs* 1 at the base of the

6

lower or both racemes, the spikelets resembling the pedicelled. *Sessile spikelets* linear-elliptic, 4–6·5 mm. long, yellowish green to violet, villous with white hairs; callus 0·5–1·5 mm. long, subacute to acute; awn 10–35 mm. long, the column puberulous with appressed hairs 0·2–0·3 mm. long. *Pedicelled spikelets* narrowly lanceolate, 3–7 mm. long, acute, muticous, white villous; pedicel tooth subulate, 0·2–1 mm. long.

ILLUSTRATIONS. Host, Gram. 4: t. 1 (1809); Pilger in Engler & Prantl, Nat. Pflanzenfam. 14e: 173, t. 92a (1940).

MINOR VARIATIONS. Two homogamous pairs at the base of the upper raceme occur very rarely, and are then associated with a fairly high number of awns per pair. 5–7 awns per raceme-pair are occasionally found on specimens which are obviously depauperate; also, in company with 1 homogamous pair on the upper raceme, in a few intermediate specimens of tropical regions. The spikelets are typically villous, but very sparsely hairy specimens may sometimes be encountered (*Milne-Redhead & Taylor* 10282 from Tanganyika seems to belong here, although its spikelets are glabrous).

The following specimens resemble *H. hirta*, except that the awn hairs are 0·7–0·9 mm. long; they are probably the result of introgression from *H. fili-pendula*. Rhodesia: Chipinga (Apr.), *Brain* 10661. Angola: Henrique de Carvalho (Apr.), *Gossweiler* 11641; Cacola (Apr.), *Gossweiler* 11766. Transvaal: Potchefstroom (Feb.), *Louw* 1682.

CHROMOSOME NUMBER. France, 2n = 30 (Larsen, 1954). Israel, 2n = 30 (Celarier, 1956). Iraq, n = ?30 (Gould, 1956). Assam, 2n = 30 (Celarier, 1956). Kenya, 2n = 30 (Krupko, 1955). South Africa, 2n = 30 (de Wet, 1954); 2n = 30 (Krupko, 1953); n = 15 (Garber, 1944); n = 20 (Gould, 1956); 2n = 44 (Krupko, 1953); 2n = 45 (de Wet, 1954). Costa Rica, n = 20 (Gould, 1956). Krupko (1953) observes that there are slight morphological differences in the leaf-blades of the two ploidy levels he encountered. Lest it be thought that the European population is less variable cytologically than the South African, it may be noted that Celarier has annotated herbarium sheets with unpublished counts of 2n = 40 for Israel, 2n = 44 for Portugal and 2n = 46 for Cyprus.

HABITAT. Dry soils in open, often overgrazed, places with a Mediterranean climate, extending into the dry tropics, particularly in highland regions.

DISTRIBUTION. Shores of the Mediterranean, extending through the Middle East to Afghanistan and Pakistan. To the west its southern limit lies in the Cape Verde Is. and Hoggar massif, but in the east it extends southwards through Egypt and Arabia to northern Tanzania. It is absent from West Africa, and uncommon in the Congo and Zambesi regions, but reappears in South Africa. Probably introduced in Australia and America (Map 5, p. 73).

FRANCE. Vernet-les-Bains (June), *Ellman & Sandwith* 261; Lauzet (July), *Proal*; Toulon (July), *Billot* 2764; Monte Carlo, La Turbie (May), *Meebold*; Mentone (Mar.), *Brenan* 352. Corsica: Bastia, *Mabille* 55; Bonifacio (June), *Kralik* 829.

SPAIN. Cordova (May), *Ellman & Hubbard* 41; Granada, La Herradura (Apr.), *Roivainen*; Almeria (Apr.), *Bourgeau* 1504; Cartagena (Apr.), *Ellman & Sandwith* 404; Barcelona, *Gonzalo* in *Sennen* 5552. MAJORCA: Palma (Apr.), F.A. *Rogers*; Alcudia (Dec.), *E. W. Kennedy* 146; Soller (May), *Bourgeau* 2812.

GIBRALTAR. Without locality, *Gamble* 28224.

PORTUGAL. Serra de Monsanto (May), *Welwitsch*.

ITALY. Noli (July), *Huet du Pavillon*; Bordighera (Apr.), *Bicknell & Pollini* in *Kneucker* 64; Ventimiglia (July), *Kuntze* in *Kneucker* 64a; Amalfi (June), *Heldreich*. SICILY: Palermo (July), *Strobl*; Taormina (Oct.), *Gamble* 20319; Catania (June), *Strobl*. SARDINIA: Santa Teresa Gallura (June), *Reverchon* 11; Laconi (July), *F. A. Muller*; Cagliari (Apr.), *Titilen* 175.

MALTA. Without locality, *Swainson, Gamble* 28141.

YUGOSLAVIA. Trau, *Pichler*; Salona, *Pichler* 1069; Split, *Petter* 64; Dubrovnik (June, Aug.), *Dennis, Turrill* 1052.

ALBANIA. Durrës (Sept.), *Alston & Sandwith* 2776; Sarandë (May), *Alston & Sandwith* 1256.

GREECE. Alexandroupolis (May), *Adamovic*; Thessalonika (Apr.), *Adamovic*; Megara (May), *Orphanides* 1193; Piraeus (May), *Heldreich* 893. CRETE: Canea (June), *Reverchon* 182; S. Nicolas, *Gandoger* 1425; Mikronisi (May), *Rechinger* 13103. AEGEAN IS: Syra, *Gandoger*; Mykonos (Apr.), *Rose Innes*; Samos (May), *Davis* 1696; Rhodes (May), *Rechinger* 7394.

TURKEY. Emick (June), *Dudley* 35445; Antalya (May), *Townsend* 63/95; Ulas to Samlar (Apr.), *Davis & Hedge* 26457; Kesliturkmenli (May), *Hennipman et al.* 1076; Iskenderum (June), *Karaplizil* 86.

CYPRUS. Episkopi (May), *Kotschy* 617; Athalassa, *Lindberg*; Larnaca (May), *E. Chapman* 692; Skala (Mar.), *Meikle* 2039.

SYRIA. Nusairy Mts. (Apr.–June), *Haradjian* 2732, 2851, 3398.

LEBANON. Broumana (May), *Mooney* 4416; Beirut (Dec.), *Hartmann* in *Kneucker* 66; Shemlan (Apr.), *Maitland* 33.

ISRAEL. Tiberias (Apr.), *Davis* 4183, 4625; Tel Aviv (Dec.), *Gabrielith* 36b.

JORDAN. Juffian (June), *Mooney* 4362; Wadi Heidan (Apr.), *Davis* 9561; Wadi Rum (May), *Gillett* 16044.

IRAQ. Darband-iBasian (June), *Haussknecht*; Bekhme Gorge (May), *Gillett* 8232; Jabal Hamrin (Mar.), *Guest* 1798; Tauq Chai (May), *Guest* 15545.

IRAN. Lorestan-Sheshom (Apr.), *Jacobs* 6415; Behbahan (Apr.), *Merton* 3789; Minab (Mar.), *Behboudi* 90326; Tchahar-bahar (Apr.), *Charif* 1184.

AFGHANISTAN. Without locality, *Griffith* 350.

SAUDI ARABIA. Dharan (Apr.), *Mandaville* 166; Anaiza (Mar.), *Vesey-FitzGerald* 15654/1; Taif (May), *Vesey-FitzGerald* 17016/1, 17030/4; J. Qurnait (May), *Vesey-FitzGerald* 17056/5; Wady Ibn Hashabil (June), *Vesey-FitzGerald* 15998/3.

BAHRAIN. *Good* 341, 342, 343, 344.

EGYPT. Alexandria (Mar.), *Ehrenberg*; Cairo, *Keller* 51; Helwan (Feb.), *Davis* 6247b; Wadi Ataqa (Mar.), *Nabil el Kadidy*; Mt. Sinai (May), *Schimper* 101.

LIBYA. Tripoli (May), *Bornmüller* 933; Garabulli (Dec.), *Park* 103; Martuba (Apr.), *Pampanini* 256; Cyrene (Apr.), *Park* 340; Cyrene to Apollonia (Apr.), *Sandwith* 2576.

TUNISIA. Nabel, *Gandoger*; Sidi Boul Baba (May), *Kralik* 311; Gabes (Jan.), *Pitard* 284.

ALGERIA. Oran (Apr.), *Debeaux*; Alger (Oct.), *Gandoger* 170; Constantine (Sept.), *Bové*.

MOROCCO. Marrakesh (May), *Trethewy* 175; Asni (May), *de Wilde & Dorgelo* 2240; Xauen (Apr.), *de Wilde & Dorgelo* 1554; Tangier (Apr.), *Lindberg* 1289.

CENTRAL SAHARA. Taharanet (Mar.), *Meinertzhagen* 211; Tamanrasset (Feb.), *Meinertzhagen* 196.

MADEIRA. Without locality, *MacGillivray* 99; *Clayton* 4040; *Grabham* 34.

CANARY Is. Palma, Caldera (June), *Sprague & Hutchinson* 353; Gomera, La Cumbrecita (Oct.), *Czeczott* 201; Tenerife, Santa Cruz (Feb.), *Borgensen* 565; Gran Canaria, Las Palmas (Apr.), *Murray.*

PAKISTAN. Dir, *Wingate*; Chitral, *Hassan ud Din* 6; Swat, Barikot (Apr.), *Ali* 26071; Swat, Mingora (Aug.), *R. R. Stewart* 24762.

AUSTRALIA. New South Wales: Coolatai (Nov.), *Johnson & Constable* NSW 30422; Warialda (July, Nov.), *Moodie, Vickery* NSW 12999.

CAPE VERDE Is. S. Vincent (June), *Vogel*; Fogo (July), *Chevalier* 44806, 44973.

CONGO REPUBLIC. Sakania to Ndola (Apr.), *Symoens* 9320; Elisabethville (Apr.), *Bequaert* 310; Keyberg, *Detilleux* 721; Muniama (Apr.), *Quarré* 7839.

SUDAN. Jebel Marra, Nyertete (Feb.), *Wickens* 1170; Jebel Marra, Digile valley (Dec.), *J. K. Jackson* 3329; Red Sea Hills, Diris Pass (Apr.), *J. K. Jackson* 2855; Suakin, *Schweinfurth* 1035; Erkowit (Mar.), *Andrews* 3554.

ERITREA. Keren (Oct.), *M. N. Harrison* 1298; Asmara (Jan.), *Pichi-Sermolli* 80; Asmara to Arbaroba (Oct.), *Scott* 206; Mt. Bizen (May), *Schweinfurth & Riva* 2016; Matcallat (Apr.), *Pappi* 4807.

ETHIOPIA. Without locality, *Schimper* 1056; Adoa (Nov.), *Schimper* 936; Addis Ababa (Nov.), *Mooney* 5061; Jimma (June), *R. B. Stewart* 71; Curfacelli (Sept.), *Burger* 983; Mt. Delo (Jan.), *Gillett* 15061.

SOMALIA. Kabalgabat (Dec.), *Gillett* 4679; Hargeisa, *Farquharson* 40; Golis Range, *Drake Brockman* 186; Daganyado (Feb.), *Glover & Gilliland* 735; Buran, *Collenette* 37.

SOCOTRA. Without locality, *Balfour* 386, *Schweinfurth* 482.

UGANDA. Moroto Mt. (July), *Liebenberg* 1748; Kamalinga (Dec.), *Brown* 21.

KENYA. Mt. Kulal (Oct.), *Bally* 5538a; Kitale (Sept.), *Napper* 783; Nakuru (Sept.), *Hitchcock* 25080; Nairobi (Feb.), *Bogdan* 2810; Narok to Kijabe (Apr.), *Glover & Samuel* 2747.

TANZANIA. Musoma, Klein's Camp (Apr.), *Greenway & Turner* 9999; Kilimanjaro (Jan.), *Haarer* 1700; Mombo to Soni (June), *Drummond & Hemsley* 3002; Mbulumbulu (July), *Greenway* 6766; Ruhudje R. (Apr.), *Schlieben* 732.

ZAMBIA. Kanona (Apr.), *Phipps & Vesey-FitzGerald* 2965.

BOTSWANA. Lobatsi (Oct.), *C. Sandwith* 99.

ANGOLA. Henrique de Carvalho (Apr.), *Gossweiler* 11664.

ASCENSION I. *Duffey* 1/63, 1/99, 2/94.

MADAGASCAR. Mandritsara (Oct.), *Bosser* 16652.

LESOTHO. Leribe (Feb.), *Phillips* 500, *Dieterlen* 203; Maluti Mts., *Staples* 104.

SOUTH WEST AFRICA. Grootfontein (Feb.), *Schoenfelder* 72; Windhoek (Mar.), *de Winter* 2579; Nauchas to Areb (Jan.), *Pearson* 9029; Bullspoort (Dec.), *Rodin* 2888; Rehoboth (Aug.), *Strey* 2164.

SOUTH AFRICA. Transvaal: Louis Trichardt (June,), *F. A. Rogers* 21169; Lydenburg (June), *Burtt-Davy* 469a; Belfast, Machadodorp (Apr.), *Pole-Evans* 3682; Pretoria (Aug.), *Moss* 5454; Johannesburg (May), *E.S.C.A. Herb.* 301; Vereeniging (Mar.), *Burtt-Davy* 17211. Orange Free State: Groot Doorukop (Dec.), *Goosens* 915; Bethlehem (Oct.), *Richardson*; Bloemfontein, *Rehmann* 3735; Fauresmith (June), *C. A. Smith* 4152. Natal: Mahlabathini (Apr.), *Gerstner* 4348; Cathedral Peak (Apr.), *Killick* 2357; Weenen, *Acocks* 11474; Maritzburg (Mar.), *McClean* 189; Durban (Apr.), *J. M. Wood* 6046. Cape

Province: Setlagodi (Apr.), *Brueckner* 373; Taungs (Feb.), *Rodin* 3624; Port St. John's (Feb.), *Moss* 5499; Clanwilliam (June), *Bolus* 14986; Capetown (Jan.), *Ecklon* 87.

MEXICO. Monterey (Feb., July), *Johnson & Barkley* 15048M, 16048M, *Pringle* 1967.

CUBA. Habana (Sept.), *Ekman* 999, 13195.

SANTO DOMINGO. Monción (June), *Ekman* 13052; Llano Costero (Feb.), *Ekman* 11419; Cuesta de Piedras to Palo (Feb.), *Ekman* 14298.

VENEZUELA. Cacute (Apr.), *Burkart* 16814.

H. hirta has a greater latitudinal distribution than any other species of the genus, and its morphological amplitude is likewise very wide. In fact one gains the impression that we are here dealing with a compilospecies, whose rapacious introgression has all but obliterated its boundaries with adjacent taxa, so that its circumscription becomes a matter of great difficulty. However it can be seen that its main area of distribution is in the Mediterranean region, a territory where no other member of the genus is found. We may therefore take this population as the model from which to draw up a circumscription of the species, free from fear of introgression which seems so obtrusive elsewhere. Even here

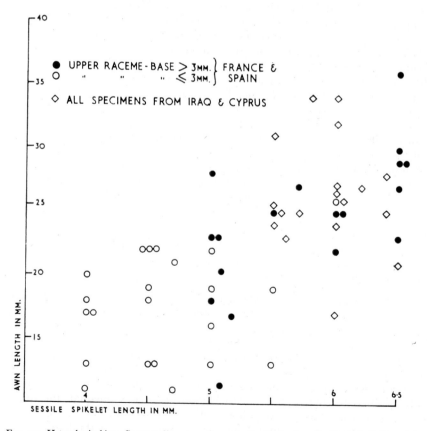

FIG. 19. *Hyparrhenia hirta*. Scatter diagram of awn and spikelet lengths for populations from eastern and western Mediterranean.

the species is not uniform, as can be seen from figure 19 (p. 81) where the compactly homogeneous population from the Middle East is compared with the much wider morphological scatter of the western Mediterranean specimens. The latter seems to be due to greater plasticity in a single population, rather than to the entry of a second taxon. It raises the uncomfortable thought that perhaps the even greater variation found in the tropics may be of a similar nature. I have chosen to ignore this possibility, since the present need is to display relationships which might be tested experimentally, and this end is best served by taking a narrow view of the species.

H. hirta is best recognized by the scanty panicle of white villous racemes which never deflex, by the harsh narrow leaves, and by the densely packed intravaginal innovations forming a hard basal tussock of culms and sheaths overlying interlocked rhizomes and roots. The differences between *H. hirta* and adjacent species may be summarized as follows:

H. quarrei. Links *H. hirta* with *H. nyassae*, differing from the former in its deflexed racemes; it also tends to have taller culms and broader coarser leaf-blades.

H. griffithii. Distinguished from *H. hirta* by the deflexed racemes and longer awns.

H. anamesa. Links *H. hirta* to *H. filipendula*, differing from the former principally in its 2 homogamous pairs on the upper raceme.

H. dregeana. Has deflexed racemes (though sometimes tardily so) with the upper raceme-base up to 1·5 mm. long (rarely more), flat and almost obscured by stiff glassy bristles. Intergrades almost imperceptibly with *H. hirta*, but fortunately the truly intermediate specimens are few in number.

H. finitima. Spikelets usually glabrous, but may bear some resemblance to the least hairy specimens of *H. hirta*. The awned pedicelled spikelet is the best distinguishing character.

This is not the place for a detailed examination of intra-specific variation, which calls for combined morphological, cytogenetic and ecological studies. Of the two classical varieties, var. *podotricha* is distinguished by the hairy summit of its peduncle, a character notoriously variable in the genus; and var. *pubescens* by its low stature and filiform leaves, a habit which can apparently be induced by heavy grazing. I therefore have no hesitation in rejecting them both as unacceptable.

13. **Hyparrhenia quarrei** *Robyns*, Fl. Agrost. Congo Belge 1: 171 (1929), & in Bull. Jard. Bot. Brux. 8: 234 (1930). Type: Congo Republic, *Quarré* 199 (BR, holotype).

Caespitose perennial; culms 1–2 m. high. *Leaf-sheaths* at base of plant usually white pubescent, sometimes glabrous, rarely tomentose; ligule 1–2 mm. long; blades rigid, up to 40 cm. long and 5 mm. wide, harsh, glaucous. *Spathate panicle* narrow, moderately dense, about 30 cm. long; spatheoles linear, 3–5 cm. long, russet coloured; peduncles a little longer than the spatheole, with spreading white hairs above. *Racemes* deflexed, though sometimes tardily and imperfectly so, 1·5–2 cm. long, 6–10-awned per pair, white hairy, terminally exserted on the flexuous peduncle; raceme-bases unequal, the upper 2–3·5 mm. long, hirsute, sometimes with a few stiff bristles, or sometimes glabrous; internodes 2–3 mm. long. *Homogamous pairs* 1 at the base of both racemes, or of the lower only, the spikelets similar to the pedicelled. *Sessile spikelets* narrowly lanceolate, 4·5–5·5 mm. long, pubescent to villous with white hairs; callus linear to slenderly cuneate, 0·7–1·2 mm. long, acute to narrowly truncate at the tip; awn 1·8–3·6 cm. long, the column pubescent with hairs 0·2–0·5 mm. long. *Pedicelled spikelets* linear lanceolate, 5–7 mm. long, acute; pedicel tooth subulate, 0·1–0·5 mm. long.

MINOR VARIATIONS. *J. N. Davies* 252 from Booué, Gabon, lies between *H. quarrei* and *H. griffithii*; its dimensions agree with the former species but it has awned pedicelled spikelets and rather long internodes.

HABITAT. Roadsides, clearings and disturbed sites.

DISTRIBUTION. Rather widely distributed in tropical Africa, but most frequent in the southern half of the continent (Map 5, p. 73). Probably introduced to Australia.

AUSTRALIA. Cowaramup (Feb.), *Becher*; Warialda (Jan.), *Vickery* 13001.
NIGERIA. Maska (Nov.), *Thatcher* S. 510; Abuja (Dec.), *Thatcher* S. 585.
CONGO REPUBLIC. Boyoko to Soro (July), *Germain* 8812; Elizabethville (Apr., May), *Gathy* 741, *Symoens* 11003; Kafubu (Mar.), *Quarré* 199.
ERITREA. Acrour (Mar.), *Schweinfurth & Riva* 1089.
KENYA. Eldoret (Aug.), *Bogdan* 1912; Thompson's Falls (Nov.), *Heady* 1277; Molo (May), *McCallum Webster* K76; Nairobi (Mar.), *Bogdan* 422; Narok, Entasekera (July), *Glover, Gwynne, Samuel & Tucker* 2062.
TANZANIA. Marangu (Sept.), *Hitchcock* 24598; Iringa Prov., Njombe, *Emson* 36; Songea (Apr.), *Milne-Redhead & Taylor* 1903; Kigonsera (Apr.), *Milne-Redhead & Taylor* 9630.
MOZAMBIQUE. Angónia (May), *Mendonça* 4185.
MALAWI. Misuku (June), *G. Jackson* 558; Ft. Manning, Nyoka (Apr.), *G. Jackson* 777; Lilongwe (Apr.), *G. Jackson* 2220.
ZAMBIA. Luwingu (May), *Astle* 609, 624, 3022; Mumbwa to Mankoya (Apr.), *Mitchell* 25/37; Choma, Mochipapa (Apr.), *Astle* 2991.
RHODESIA. Poole (Mar.), *Hornby* 2322; Mandoro Reserve (June), *R. M. Davies* GH 32694, GH 32696; Marandellas (May), *A. Newton* GH 20943.
ANGOLA. General Machado (May), *Gossweiler* 11288.
SOUTH AFRICA. Transvaal: Lydenburg (June), *Burtt-Davy* 469; Buffelspan (Dec.), *Pole-Evans* 648; Pretoria (Feb.), *Pole-Evans* 424; Johannesburg (Apr.), *Moss* 6853; Piet Retief to Wakkerstroom (May), *Pole-Evans* 3699. Orange Free State: Memel (Apr.), *Acocks* 12563. Natal: Newcastle (May), *Pole-Evans* 3784; Ladysmith (Feb.), *Godfrey* 1895; Cathedral Peak (Apr.), *Killick* 2358; Estcourt (Apr., May), *West* 765, *Acocks* 11496. Cape Province: Hunterstown (Mar.), *Acocks* 15724; Windvoelberg (Apr.), *Roberts* 1745.

H. quarrei is what I have termed a linking species, with specimens at either end of its range of variation merging into *H. hirta* and *H. nyassae* respectively. It differs from the former in its deflexed racemes, and from the latter in its white indumentum. It also tends to intergrade with *H. tamba* and *H. dregeana*, but whereas these species have a flat upper raceme-base almost obscured by stiff bristles and rarely over 1·5 mm. long, that of *H. quarrei* is rather longer, clearly terete, and sparsely, if at all, bristly.

The predominantly southern African distribution is in accord with the supposition that it may comprise introgression products between *H. hirta* and *H. nyassae*, a hypothesis marred by the anomalous Nigerian specimens.

14. **Hyparrhenia griffithii** *Bor* in Ind. For. Rec. (Bot.) 1 : 92 (1938). Type: India, *Griffith* 6766 (K, holotype).

Andropogon griffithii Munro in Cat. Pl. Griffith: 58 (1865), *nom. nud.* Based on *Griffith* 6766.

Caespitose perennial; culms 1–2 m. high. *Leaf-sheaths* glabrous or sparsely pilose; ligule 2 mm. long; blades up to 40 cm. long and 8 mm. wide. *Spathate panicle* narrow, lax, up to 35 cm. long; spatheoles linear, 4–7 cm. long, convolute; peduncles flexuous, from ⅔ to rather longer than the spatheole, pilose with white hairs towards the top. *Racemes* strongly deflexed, 1·5–3·5 cm. long, 5–10-awned per pair, loose, the internodes 3·5–4·5 mm. long; raceme-bases unequal, the upper 3·5–8 mm. long, glabrous. *Homogamous pairs* 1 at the base of both racemes, the spikelets similar to the pedicelled. *Sessile spikelets* lanceolate, 6–7 mm. long; callus pungent 1·5–2 mm. long; lower glume brownish to dark violet, hirsute with white hairs; awn 4–6 cm. long, the column pubescent with rufous hairs 0·4–0·6 mm. long. *Pedicelled spikelets* linear-lanceolate, 6–8 mm. long, villous, with or without an awnlet up to 4 mm. long; pedicel tooth subulate, 0·5 mm. long.

MINOR VARIATIONS. Very occasionally 2 homogamous pairs occur on the upper raceme of a few of the raceme-pairs in a panicle.

HABITAT. A species vaguely described as 'in grassland'.

DISTRIBUTION. Abundant but very localized in India; also occurring in Madagascar and scattered localities in Africa (Map 6, p. 76).

INDIA. Assam, Khasi Hills (May, Nov.), *Griffith* 6766, *Deka* 20117, *Chand* 4650.
SUDAN. Jebel Marra, Gollol (June), *Blair* 59.
KENYA. Chyulu South (June), *Bally* 8120.
ZAMBIA. Mporokoso (Apr.), *Phipps & Vesey-FitzGerald* 3128.
MADAGASCAR. Faratsiho (Feb.), *Bosser* 10782.

H. quarrei is a somewhat heterogeneous species bounded by, and uniting, *H. hirta*, *H. nyassae* and *H. tamba*. The heterogeneity is increased by a small number of specimens with unusually long awns, sessile spikelets, rhachis internodes and raceme-bases (Fig. 20), which apparently bridge the gap between *H. quarrei* and *H. familiaris*. The latter is usually easy to recognise by its glabrous spikelets, yellow peduncle hairs, and twin homogamous pairs on the upper raceme, but there are several marginal specimens (see under minor variations of *H. familiaris*).

At the time of its original description, *H. griffithii* was considered to be endemic to India, but it is now apparent that it is identical to the long-awned variant of *H. quarrei* in Africa. Its continued acceptance as a species thus becomes open to question, but I have preferred to take a narrow view of species in this admittedly marginal case. It is perhaps analogous to *H. rufa* var. *siamensis*, in that only a fragment of the African complex (a single apomict?) seems to have migrated overseas, but it is hard to explain why it should be restricted to the Khasi Hills where it is quite abundant.

The species also merges into *H. nyassae*. The Madagascan specimen cited has white pilose racemes, but I have since seen a duplicate of this number in which many of the spikelets bear pale yellow hairs. This is presumably the

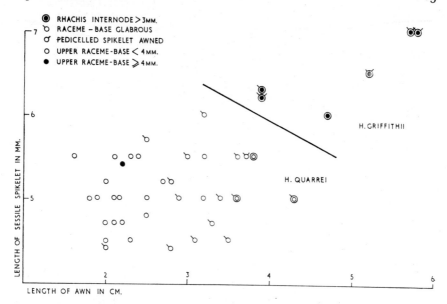

FIG. 20. Scatter diagram of characters from *Hyparrhenia quarrei* and *H. griffithii*.

result of introgression, in which case the equivocal identity of some specimens is understandable.

15. **Hyparrhenia anamesa** *W. D. Clayton*, sp. nov.; affinis *H. hirtae* (L.) Stapf, sed paribus spicularum homogamarum racemi superioris duabus et pari racemorum 4–6(–7) aristata differt; a *H. filipendula* (Hochst.) Stapf pilis columnae aristae brevioribus distinguenda. Type: Kenya, *Glover, Gwynne, Samuel & Tucker* 2145 (K, holotype).

Densely caespitose perennial; culms 60–120 cm. high. *Leaf-sheaths* glabrous or sometimes loosely hirsute; ligule 2 mm. long; blades up to 40 cm. long and 4 mm. wide but often much shorter, harsh, glaucous. *Spathate panicle* loose, 15–45 cm. long; spatheoles linear, 4–6 cm. long; peduncles slender and flexuous, usually exceeding the spatheole, with or sometimes without spreading white hairs above. *Racemes* not deflexed, 1·5–2·5 cm. long, 4–6(–7)-awned per pair, villous with white hairs, terminally exserted; raceme-bases very unequal, the upper 3·5–6 mm. long, glabrous or sometimes softly hirsute. *Homogamous pairs* 1 at the base of the lower raceme and 2 at the base of the upper; spikelets linear-lanceolate, 5–6 mm. long, villous. *Sessile spikelets* linear-oblong, 5–6·5 mm. long, white villous; callus 1–1·8 mm. long, acute; awn 2·5–4 cm. long, the column pubescent with hairs 0·1–0·6 mm. long. *Pedicelled spikelets* linear-lanceolate, 4–6 mm. long, with or without an awn-point 1–2 mm. long; pedicel tooth triangular, 0·6 mm. long.

ILLUSTRATION. Fig. 21 (p. 86).

HABITAT. Dry soils and open places.

DISTRIBUTION. Eastern Africa from Sudan to the Cape (Map. 6, p. 76).

CONGO REPUBLIC. Elisabethville (Apr.), *Gathy* 649.

FIG. 21. *Hyparrhenia anamesa.* **1,** base of plant showing habit of stem, ×⅔; **2,** inflorescence, ×⅔; **3,** diagram of raceme-pair, ×3; **4,** base of raceme-pair showing homogamous spikelets and one heterogamous pair, ×6; **5,** lower glume of sessile spikelet showing callus, ×6. All drawn from *Bogdan* 2197.

Sudan. Jebel Marra, Gur Lambang (June), *Wickens* 2991.

Ethiopia. Debra Marcos (Sept.), *Hillier* 904; Addis Allem (Sept.), *Omer-Cooper*; Jimma (Oct.), *Mooney* 6029; Harar to Djigdjigga (July), *Burger &
Getahun* 361; Sidamo, Yavello (Jan.), *Mooney* 5511.

Kenya. Lodwar (Sept.), *Paulo* 998; Eldoret (Sept.), *Hitchcock* 25006; Meru
(Dec.), *Bogdan* 2693; Narok, Olodungoro (July), *Glover, Gwynne, Samuel &
Tucker* 2145; Nairobi (Feb.), *Bogdan* 2811; Coast, Matuga (Jan.), *Moomaw*
1176.

Tanzania. Kilimanjaro (July), *Greenway* 6726; Mt. Hanang (Feb.),
Greenway 7744; Lolkisale (Mar.), *Welch* 497; Kondoa-Irangi, Kolo (Feb.),
B. D. Burtt 1297; Songea (May), *Boaler* 599.

Mozambique. Massangulo (Apr.), *Sousa* 1363; Beira (Feb.), *F. A. Rogers*
5945.

Malawi. Nyika (June), *G. Jackson* 525, 855; Mvai (June), *Wiehe* N590.

Zambia. Abercorn (Apr.), *McCallum Webster* A357; Ndola (Feb.), *Fanshawe*
2868; Mankoya (Apr.), *Verboom* 1366; Kapushi R. (June), *Trapnell* 2047;
Mumbwa (June), *Mitchell* 25/38.

Rhodesia. Salisbury (Apr.), *Eyles* 2174; Marandellas (Feb.), *Corby* 387;
Inyanga (Mar.), *Gilliland* 1709; Melsetter (June), *Crook* 479; Nyamandhlova
(Apr.), *Plowes* 1618; Matopos (Feb.), *Rattray* 1590.

Lesotho. Maseru (Jan.), *Liebenberg* 5638.

South Africa. Transvaal: Louis Trichardt (Mar.), *van Vuuren* 1684;
Pietersburg (Jan.), *van Vuuren* 1360; Pilansberg (Dec.), *Pole-Evans* 644;
Krugersdorp (Feb.), *Rodin* 3821; Pretoria (Feb.), *Pole-Evans* 14; Delmas to
Bethal (May), *Pole-Evans* 3700. Orange Free State: W. Pretorius Game Reserve
(Feb.), *van Zinderen Bakker* 1068. Natal: Dargle (Feb.), *Godfrey* 1916; Krantz-
kloof (Mar.), *Kuntze*; Weenen (May), *Acocks* 11495; Underberg (Mar.),
McClean 634; Riet Vlei, *Buchanan*. Cape: Table Mt., *Drege*; Stellenbosch
(Nov.), *Godfrey* 1277; George, Kaaiman's R. (Mar.), *Wilman* 25479; Grahams-
town, *Macowan* 494; Keiskama (Aug.), *Story* 3888.

In the discussion of *H. filipendula* it is shown that *H. hirta*, as judged by its
Mediterranean population, and *H. filipendula* are distinct species, but that in
the tropics they are linked by intermediate specimens (figs. 23 & 24, p. 100 &
101). It is also shown that the hirtellous awns of *H. filipendula* enable it to be
separated from the complex. Can we, in a like manner, define the limits of
H. hirta and thus isolate the intermediate specimens?

Unfortunately there seems to be no simple distinction, for variation throughout the complex is continuous. It may be noted however that among the
Mediterranean plants 2 homogamous pairs at the base of the upper raceme are
extremely rare, and that the number of awns per raceme-pair falls below 8
only in a few obviously depauperate specimens. In the tropics the situation is
different; plants of the usual Mediterranean type are found but also a high
proportion of plants with fewer than 8 awns per pair, most of which also have 2
homogamous pairs on the upper raceme. The concurrence of both these
characters provides us with the means, of necessity somewhat arbitrary, for
separating the intermediate plants from the typical, or Mediterranean, population of *H. hirta*. They are here placed in a new species, which may possibly have
arisen as a result of hybridization between *H. hirta* and *H. filipendula*.

This solution works well enough north of the equator, where the geographical
distribution of the two species is obviously different. In South Africa it is less

convincing for the two species are intermingled to a much greater degree, and have the appearance of a single very variable population. This is perhaps to be expected in a region where tropical and Mediterranean climatic regimes are in close juxtaposition.

Our species has sometimes been confused with *H. buchananii*. The type of the latter however has rufously hairy spikelets and should be identified with *H. poecilotricha*.

16. **Hyparrhenia violascens** (*Stapf*) *W. D. Clayton*, stat. nov.

H. soluta var. *violascens* Stapf in Prain, Fl. Trop. Afr. 9: 319 (1918). Type: Nigeria, *Dalziel* 263 (K, holotype).

Annual; culms erect, 1 m. high. *Leaf-sheaths* glabrous; ligule 1 mm. long; blades up to 40 cm. long and 6 mm. wide, the collar glabrous or with a few hairs. *Spathate panicle* lax, scanty, 20–30 cm. long; spatheoles linear, 4–7 cm. long; peduncles flexuously exserted from the tip of the spatheole, with spreading white hairs near the top. *Racemes* tardily deflexed, 1·5–3 cm. long, 6–9-awned per pair, yellowish white strongly tinged with violet; raceme-bases unequal, the upper 2–3 mm. long, glabrous or with a few hairs, the foot of the

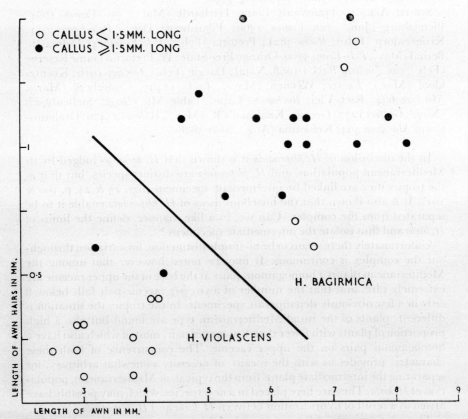

FIG. 22. *Hyparrhenia violascens* and *H. bagirmica*. Length of awn plotted against length of awn hairs and callus.

bases glabrous. *Homogamous pairs* 1 at the base of the lower raceme only; spikelets narrowly lanceolate, 7–8 mm. long, pubescent with white or some- times fulvous hairs. *Sessile spikelets* linear-lanceolate, 5–6·5 mm. long, pubescent to villous with white or off-white hairs (rarely almost glabrous); callus slender, 0·8–1·2 mm. long, acute to narrowly truncate; awn 2·5–4 cm. long, the column pubescent with white or fulvous hairs 0·1–0·5 mm. long. *Pedicelled spikelets* linear-lanceolate, 6–7 mm. long, acute to acuminate, pubescent with white or yellow hairs; pedicel tooth linear, 0·3–0·7 mm. long.

MINOR VARIATIONS. There is some variation in the colouring of the homo- gamous and pedicelled spikelets, but the indumentum of the sessile spikelets is always white.

HABITAT. A species of roadsides and old cultivation.

DISTRIBUTION. The Sudan vegetation zone of West Africa (Map 6, p. 76).

NIGERIA. Sokoto, Bimasa Bakuro-Tureta F.R. (Nov.), *Latilo* FHI 43780; Katagum, *Dalziel* 263; Fika (Oct.), *Daggash* FHI 24880; Shendam, Lowlands Farm (Nov.), *Clayton* 1469; Abinsi (Nov.), *Dalziel* 891; Vogel Peak, Gurum (Nov.), *Hepper* 1282.

CAMEROUN. Maroua (Oct.), *Vaillant* 2477; Garoua (Oct.), *Koechlin* 7349. CHAD. Bousso to Fort Archambault (Nov.), *Chevalier* 10490 bis.

H. violascens closely resembles *H. quarrei* in all but its annual habit and glabrous basal sheaths; less reliable differentiating characters are the glabrous foot of the raceme-bases, and absence of coarse hairs at the collar of the leaf.

On present evidence there is sufficient disjunction between *H. violascens* and *H. bagirmica* to justify regarding them as distinct species (Fig. 22). How- ever the number of herbarium specimens available is rather small, and their division into two size classes may be purely fortuitous.

17. **Hyparrhenia bagirmica** (*Stapf*) *Stapf* in Prain, Fl. Trop. Afr. 9: 319 (1918).

Cymbopogon bagirmicus Stapf in Journ. de Bot., sér. 2, 2: 214 (1909). Type: Chad, *Chevalier* 9795 (P, lectotype; K, isotype).
C. solutus Stapf, *l. c.*: 211 (1909). Type: Chad, *Chevalier* 10507 (P, lectotype).
Andropogon bagirmicus (Stapf) Chev., Sudania 1: 166 (1911).
A. brachypodus Stapf ex Chev., *l. c.*: 167 (1911), *nom. nud.* Based on 2 specimens from Chad (*Chevalier* 9849, 9850 (both P)).
Hyparrhenia soluta (Stapf) Stapf in Prain, Fl. Trop. Afr. 9: 318 (1918).

Annual; culms erect, up to 2 m. high. *Leaf-sheaths* glabrous at the base; ligule 1·5–3 mm. long; blades up to about 30 cm. long and 3 mm. wide. *Spathate panicle* lax, narrow, 20–80 cm. long; spatheoles linear, 5–8 cm. long, reddish; peduncles about as long as the spatheole, with spreading white or yellow hairs towards the tip. *Racemes* deflexed, 1·5–2·5 cm. long, 6–9-awned per pair, terminally exserted, pale yellow or ashen tinged with violet; raceme- bases unequal, the upper 2–3 mm. long, glabrous or with a few hairs, the foot of the bases glabrous. *Homogamous pairs* 1 at the base of the lower raceme only; spikelets linear-lanceolate, 7–8 mm. long, pubescent with white or fulvous hairs. *Sessile spikelets* linear-lanceolate, 6–7 mm. long, pubescent with white hairs; callus pungent, 1·5–2·5 mm. long; awn 5–8 cm. long, the column dark brown to black and hirtellous with white or fulvous hairs 1–1·5 mm. long.

Pedicelled spikelets linear-lanceolate, 7–8 mm. long, pubescent with white or fulvous hairs, the lower glume acuminate or with a short mucro up to 1 mm. long; pedicel tooth linear, 0·3–0·7 mm. long.

MINOR VARIATIONS. The colour of the racemes varies from white to pale golden, according to the indumentum colour of the homogamous and pedicelled spikelets. However the indumentum of the sessile spikelets is always white.

HABITAT. A species of roadsides and old cultivation.

DISTRIBUTION. The Sudan vegetation zone of West Africa (Map 6, p. 76).

SENEGAL. Niokolo Koba (Oct.), *Adam* 15886; Tambacounda (Oct.), *Adam* 12720; Samba to Torodo (Oct.), *Adam* 18456.

NIGERIA. Katsina, Moshi (Oct.), *Clayton* 1379; Zaria, Samaru (Oct., Nov.), *Thatcher* S. 497, S. 499, S. 524, S. 526; Damagum (Oct.), *de Leeuw* 1254.

CAMEROUN. Waza (Oct.), *Koechlin* 7401; Guetalé to Waza (Oct.), *Koechlin* 7382.

CHAD. Dar el Hadjer (Sept.), *Chevalier* 9795; L. Fittri (Sept.), *Chevalier* 9850; Bousso to Fort Archambault (Nov.), *Chevalier* 10507; Massakory (Oct.), *Koechlin* 9844.

CENTRAL AFRICAN REPUBLIC. Bia (Sept.), *Audru & Boudet* 2084.

H. bagirmica is an annual species similar to *H. violascens*. It can be distinguished from both the latter and from the related *H. nyassae* by the longer awns and callus, and by the hirtellous indumentum of the awns. It was based on 4 syntypes, two of which subsequently served as a basis for *A. brachypodus*, one of Stapf's provisional names published by Chevalier as a *nomen nudum*. One of the remaining specimens has been chosen as a lectotype.

Cymbopogon solutus was based on two specimens, one of which is a complete plant of *H. bagirmica*. The other, consisting of an inflorescence only, is apparently *H. violascens*, but since the annual habit cannot be confirmed its identity will always be open to some doubt. The situation has been further confused by the addition of forma *trichophyllus*, for this is identical with *H. nyassae* and it is to specimens of the latter species that the name *H. soluta* has commonly been applied. It is perhaps fortunate that *H. soluta* can now be reduced to synonymy.

18. **Hyparrhenia barteri** (*Hack.*) *Stapf* in Prain, Fl. Trop. Afr. 9: 321 (1918).

Andropogon barteri Hack. in Flora 68: 124 (1885), & in DC., Monogr. Phan. 6: 635 (1889). Type: Nigeria, *Barter* (K, isotype).
Sorghum barteri (Hack.) Kuntze, Rev. Gen. Pl. 2: 791 (1891).

Annual; culms up to 2 m. high. *Leaf-sheaths* glabrous; ligule 1 mm. long; blades up to 30 cm. long and 4 mm. wide, scaberulous. *Spathate panicle* narrow, dense, 30–40 cm. long, of up to 8 fastigiate tiers one above the other; spatheoles linear, 3–4 cm. long, at length reddish-brown, embracing the base of the racemes; peduncles up to half as long as the spatheoles, with or without spreading white hairs near the tip. *Racemes* not deflexed, 8–10 mm. long, 2-awned per pair, greenish or pallid, laterally exserted; raceme-bases unequal, the upper 4–7 mm. long, glabrous. *Homogamous pairs* 1 at the base of each raceme; spikelets narrowly lanceolate, 3–5 mm. long. *Sessile spikelets* narrowly lanceolate, 5·5–6·5 mm. long; callus pungent, 1·5–2 mm. long or rarely longer; lower glume glabrous, 9-nerved, the inner 3–5 nerves more or less raised and scaberulous; awn 4–4·5 cm. long, the column coarsely hirsute with rufous hairs

3–5 mm. long. *Pedicelled spikelets* linear-lanceolate, 3·5–5 mm. long, glabrous, terminating in an awn-point 1 mm. long; pedicel tooth subulate, 0·1 mm. long.

MINOR VARIATIONS. Some of the dimensions are occasionally slightly longer than above, such as awn 5 cm., callus nearly 3 mm., and pedicelled spikelet awn 3 mm. Although approaching the measurements of *H. figariana*, they seem to occur as sporadic variants rather than true intermediates.

Gérard 1913 from Niangara in the Congo Republic is curious in that the awn hairs are mostly short (0·5 mm. long), but they are interspersed with a few longer hairs (up to 2 mm. long) appressed to the column of the awn.

HABITAT. A species of old farmland and roadsides on poor soils.

DISTRIBUTION. Around the periphery of the Congo basin from Togo to Zambia (Map 7).

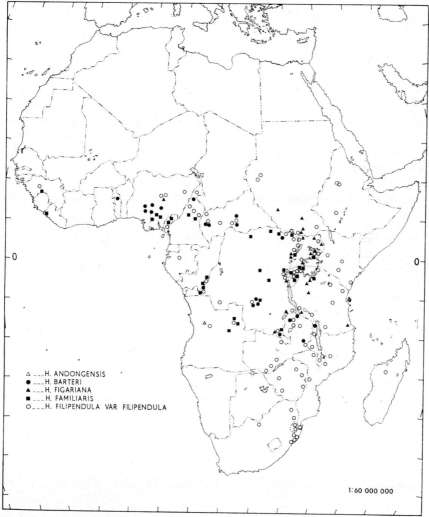

MAP 7. Distribution of *Hyparrhenia andongensis, H. barteri, H. figariana, H. familiaris* and *H. fili-pendula* var. *filipendula*.

TOGO. Sokode to Lama Kara (Oct.), *Rose Innes* GC 31391.

N. NIGERIA. Wamba, Fadau Karshi (Oct.), *J. D. Kennedy* FHI 8037; Niger Prov. (Nov.), *Freeman* S. 196; Gwada (Sept.), *Lamb* FHI 3159; Lokoja (Oct.), *Dalziel* 295, *Barter*; Okene (Oct.), *J. F. Ward* L. 145

S. NIGERIA. Awka (Oct.), *N. W. Thomas, A. P. D. Jones* FHI 6791; Ezillo (Oct.), *Tuley* 933.

CAMEROUN. Garoua to Kapsiki (Oct.), *Koechlin* 7354; Rônicis valley (Oct.), *Koechlin* 7290.

CENTRAL AFRICAN REPUBLIC. Kaga Do (Oct.), *Chevalier* 5925, 5948; Gomoko (Sept.), *Audru & Boudet* 1895; Bouar (Sept.), *Koechlin* 6246; Kou-kourou (Sept.), *Koechlin* 10644.

CONGO REPUBLIC. Lukusa (Sept.), *Liben* 3694; Katabaie (May), *Liben* 2931; Gandajika (Apr.), *Risopoulos* 265; Dungu (Sept.), *Gérard* 674; Musosa (Nov.), *Bredo* 181.

MALAWI. Karonga (Mar.), *G. Jackson* 1257.

ZAMBIA. Mpulungu (Apr.), *Phipps & Vesey-FitzGerald* 3010; Musesha to Muzombwe (Apr.), *Phipps & Vesey-FitzGerald* 3183; Machinje hills (Mar.), *Astle* 4620, 4629.

An annual species clearly characterized by the remarkable long hairs on the column of the awn. It differs from other members of the *Filipendula* group in having a single homogamous pair at the base of the upper raceme, and a shorter upper raceme base. It can usually be recognized at sight from the appearance of the panicle, in which successive tiers do not overlap, but each forms a tight bundle of spathes, spatheoles and racemes bunched around the axis.

19. **Hyparrhenia familiaris** (*Steud.*) *Stapf* in Prain, Fl. Trop. Afr. 9: 325 (1918).

Andropogon familiaris Steud., Syn. Pl. Glum. 1: 385 (1854); Hack. in DC., Monogr. Phan. 6: 636 (1889). Type: 'Guinea', *Jardin* (P, holotype).

Themeda effusa Bal. in Journ. de Bot. 4: 115 (1890). Type: Vietnam, *Balansa* 1726, 1727 (both P, syntypes; and K, isosyntypes).

Sorghum familiare (Steud.) Kuntze, Rev. Gen. Pl. 2: 791 (1891).

Andropogon kiwuensis Pilger in R. E. Fries, Wiss. Ergebn. Schwed. Rhod.-Kongo-Exped. 1911–12, 1: 196 (1916). Type: Rwanda, *Fries* 1544 (UPS, holotype).

Anthistiria balansae Crev. & Lem., Cat. Prod. Indochine 1: 362 (1917). Based on *Themeda effusa* Bal.

Andropogon lugugaensis var. *levervilleensis* Vanderyst in Bull. Agric. Congo Belge 9: 241 (1918), *nom. prov.* Type: Congo Republic, *Vanderyst* s.n. (BR, holotype).

A. familiaris var. *levervilleensis* Vanderyst, *l. c.* (1918), *nom. prov.*, & in Bull. Soc. Bot. Belge 55: 43 (1923), in synon. sub *H. finitima.*

Cymbopogon effusus (Bal.) A. Camus in Bull. Mus. Hist. Nat. Paris 24: 536 (1918).

C. familiaris (Steud.) De Wild. in Bull. Jard. Bot. Brux. 6: 12 (1919).

Hyparrhenia effusa (Bal.) A. Camus in Bull. Mus. Hist. Nat. Paris, sér. 2, 3: 760 (1931).

Caespitose perennial; culms 50–150 cm. high, slender, glabrous. *Leaf-sheaths* glabrous or loosely pilose on the margins; ligule 1·5 mm. long; blades up to about 30 cm. long and 5 mm. wide. *Spathate panicle* leafy, lax, up to 60 cm. long; spatheoles linear, 6–10 cm. long, reddish brown; peduncles as long as or rather shorter than the spatheoles, bearing a beard of spreading yellow hairs towards the tip. *Racemes* deflexed, 1–2 cm. long, 3–5-awned per pair, olive green tinged with purple, laterally exserted on the flexuous peduncle; raceme-bases unequal, pubescent or fulvously bearded in the fork, the upper 4–5 mm. long, glabrous. *Homogamous pairs* 1 at the base of the lower, and 2 at the base of the upper raceme; spikelets linear-lanceolate, 6–6·5 mm. long. *Sessile spikelets* linear-oblong, 7·5–8·5 mm. long, glabrous or pubescent with white hairs; callus pungent, 2·5–3 mm. long; awns 6–9 cm. long, the column hirtellous with fulvous hairs up to 1 mm. long. *Pedicelled spikelets* linear-lanceolate, 6–9 mm. long, glabrous, terminating in an awn 2–10 mm. long; pedicel tooth linear, 0·5 mm. long.

ILLUSTRATION. De Wild. in Bull. Soc. Bot. Belge 51 : 256, t. 63 (1913).

MINOR VARIATIONS. The peduncle hairs are usually bright yellow, but may be pallid, or even white in a very few specimens (such as *Vanderyst* 5211 from the Congo). Glabrous spikelets are the rule, but they are occasionally white pubescent.

HABITAT. A species of savanna grasslands.

DISTRIBUTION. A species of the Congo and adjacent regions, but also occurring in Vietnam (Map 7, p. 91).

VIETNAM. Tonkin, Mt. Bovi (May, Dec.), *Balansa* 1726, 1727.

GUINEA. Mamou (Oct.), *Adam* 12755; Nzérékoré to Beyla (May), *Adam* 5050a.

LIBERIA. Robertsfield (June), *Adam* 21346.

GHANA. Amedzofe (Dec.), *Rose Innes* GC 31177.

NIGERIA. Ogoja Prov., *Rosevear* 10/30a; Obudu plateau (Nov.), *Tuley* 775, 1030.

CAMEROUN. Bamenda (Dec.), *Baldwin* 13835; Nyen (May), *Brunt* 1126; Bali (Apr.), *Brunt* 1057; Ngaoundéré (Oct.), *Breteler* 526; Meiganga to Ngaundéré (Oct.), *Koechlin* 7195, 7210, 7211; Kogan (Oct.), *Koechlin* 7502.

CENTRAL AFRICAN REPUBLIC. Goubali (Oct.), *Audru & Boudet* 2239; Djibo (Dec.), *Boudet & Bille* 1446; Bouar (Sept.), *Koechlin* 6239; Yaloké (Oct.), *Koechlin* 3033.

REPUBLIC OF CONGO. Brazzaville (July), *Chevalier* 27285; Mbé (Feb.), *Koechlin* 2582; Banza to Poundi (Mar.), *Koechlin* 3680.

CONGO REPUBLIC. Kisantu, Mawanga (Apr.), *Callens* 4052; Stanleypool (Apr.), *Vanderyst* 3780; Bumbuli (Nov.), *Lebrun* 6566; Parc Nat. Garamba (Oct.), *de Saeger* 1471; Gety (Aug.), *H. B. Johnston* 1086a; Kivu, Katana (May), *Léonard* 4244; Lomami (Sept.), *Quarré* 2666.

RWANDA. Nyanza (Jan.), *Schantz & Turner* 4228; Ruzizi valley (Dec.), *Fries* 1544.

BURUNDI. Karuzi (Apr.), *Van der Ben* 2009.

UGANDA. Buruma, *Maitland*; Kigungu (Apr.), *Hacker* 828; Entebbe (May), *Fiennes* 3374.

TANZANIA. Ngara, Kibanga (May), *Watkins* 415; Kiamawa F.R. (Nov.), *Procter* 747; Bukoba (Jan.), *Procter* 1112; Karoro (Jan.), *Haarer* 2453; Nyakato (June), *Gillman* 54.

ANGOLA. Henrique de Carvalho (Apr.), *Gossweiler* 11498, 11647; Dala (Apr.), *Gossweiler* 11224; Muriege (Apr.), *Gossweiler* 11710.

A distinctive species, differing from other members of the *Filipendula* group in its deflexed racemes, 3–5 awns per pair of racemes and more open panicle. The most easily observed feature is the beard of spreading yellow hairs at the top of the peduncle; the long awn on the lower glume of the pedicelled spikelet is a further aid to identification.

It merges with *H. poecilotricha*, and it seems reasonable to confine our species to the specimens with glabrous or white haired spikelets. It is apparent from the paratypes that the name *H. familiaris* var. *pilosa* Robyns was intended for the latter, but unfortunately the spikelets of the holotype are rufously hairy.

20. **Hyparrhenia figariana** (*Chiov.*) *W. D. Clayton*, comb. nov.

Andropogon filipendulus var. *calvescens* Hack. in DC., Monogr. Phan. 6: 635 (1889). Type: Sudan, *Schweinfurth* 2421 (K, lectotype).
Cymbopogon figarianus Chiov. in Bull. Soc. Bot. Ital. 1917: 59 (1917). Type: Sudan, *Figari*.
Andropogon figarianus Chiov., *l. c.* (1917), *in synon.*
Hyparrhenia barteri var. *calvescens* (Hack.) Stapf in Prain, Fl. Trop. Afr. 9: 322 (1918).

Annual; culms erect, 60 cm. to 2 m. high. *Leaf-sheaths* glabrous or sparsely pilose on the margins; ligule 2 mm. long; leaf-blades up to 30 cm. long and 4 mm. wide, scaberulous. *Spathate panicle* narrow, 20–60 cm. long, 2–5-tiered, usually with the successive tiers densely fastigiate; spatheoles linear, 3–7 cm. long, at length reddish brown; peduncles up to half as long as the spatheole, with, or sometimes without, spreading white hairs towards the tip. *Racemes* not deflexed, 1–1·2 cm. long, 2-awned per pair, yellowish green, laterally exserted; raceme-bases unequal, the upper 6·5–10 mm. long, glabrous. *Homogamous pairs* 1 at the base of the lower, and 2 at the base of the upper raceme; spikelets linear-lanceolate, 5–8 mm. long. *Sessile spikelets* narrowly lanceolate, 6·5–10 mm. long; callus pungent, 2–3 mm. long; lower glume glabrous, 9-nerved with the inner nerves more or less raised, the back rounded or sometimes obscurely hollowed along the median line; awns (4–)5–7·5 cm. long, the column hirtellous with fulvous hairs 0·7–1 mm. long. *Pedicelled spikelets* linear-lanceolate, 6–7 mm. long, glabrous, terminating in a short bristle 1–4 mm. long; pedicel tooth acute, 0·2 mm. long.

MINOR VARIATIONS. Although normally glabrous, the spikelets are very rarely shortly pilose with white (*H. B. Johnston* 1210 from the Congo) or pallidly fulvous hairs (*Myers* 12028 & *J. K. Jackson* 870 from the Sudan). The annual habit and 2-awned raceme-pairs scarcely permit the transfer of these specimens to any other species.

HABITAT. A species of waste land and roadsides.

DISTRIBUTION. Central east Africa from the Sudan to Tanzania, and in Nigeria (Map. 7, p. 91).

NIGERIA. Jos, Panshanu (Aug.), *Lawlor & Hall* 183, 339.
CONGO REPUBLIC. Kasenyi (Aug.), *H. B. Johnston* 1064; Semliki valley (Oct.), *H. B. Johnston* 1213; Uvira (Aug.), *Symoens* 1130; Piveto (Apr.), *Robyns* 1972; Elisabethville to Sampwe (May), *Bredo* 2852.

SUDAN. Wau (Oct.), *J. K. Jackson* 3994; Seriba Ghattas (Sept.), *Schweinfurth* 2421; Bor (Oct.), *M. N. Harrison* 1036; Mongalla, *J. K. Jackson* 320; Juba (Feb.), *Simpson* 7552.

UGANDA. Moyo (Dec.), *M. N. Harrison* 546; Madi (Nov.), *A. S. Thomas* 4059; Moroto (Aug.), *Clayton* 85; Kumi (Aug.), *Langdale-Brown* 2330; Entebbe, *Maitland*.

TANZANIA. Mwanga (Apr.), *Greenway* 7370; Kahama (May), *Welch* 81; Tabora, L. Chaya (May), *van Rensburg* 327; Kasanga (Mar.), *McCallum Webster* T227; Lindi, Lutamba (Apr.), *Schlieben* 6263.

The similar panicle shape and annual habit led Stapf (1918) to regard *H. figariana* as a variety of *H. barteri*, but the two species have little else in common and are easily distinguished.

Of more importance is Hackel's (1889) earlier suggestion that *H. figariana* should be subordinate to *H. filipendula*, for these two species are very difficult to separate without their basal parts. Certainly the false panicle of *H. figariana* tends to be contracted about the axis and broken up into successive tiers, recalling the habit of *H. barteri*; also the dimensions of spikelets, awns and other parts are, on average, longer than those of *H. filipendula*, often giving the panicle a coarser appearance. However there is so much overlap that these characters cannot be regarded as reliable. Nevertheless the difference in habit is unmistakeable, the strictly annual habit of *H. figariana* contrasting strongly with the scaly perennating rhizomes of *H. filipendula*. There is thus ample justification for regarding the former as a distinct species. We might compare the situation with that in *H. rufa*, where the difference between the annual and perennial habits is indistinct, and has been discarded as a diagnostic character.

I have been unable to locate the type of *H. figariana*, but it can be identified from the very full description, despite the unusual reduction of the homogamous spikelets of the upper raceme to a single pair.

21. **Hyparrhenia filipendula** (*Hochst.*) *Stapf* in Prain, Fl. Trop. Afr. 9: 323 (1918).

Andropogon filipendulus Hochst. in Flora 29: 115 (1846); Hack. in DC., Monogr. Phan. 6: 634 (1889); Stapf in Thistleton-Dyer, Fl. Cap. 7: 362 (1898). Type: South Africa, *Krauss* 28 (K, isotype).
A. filipendulinus Hochst. ex Steud., Syn. Pl. Glum. 1: 389 (1854), *sphalm.*
Sorghum filipendulum (Hochst.) Kuntze, Rev. Gen. Pl. 2: 791 (1891).
Cymbopogon filipendulus (Hochst.) Rendle, Cat. Afr. Pl. Welw. 2: 157 (1899).
C. filipendulus var. *angolensis* Rendle, *l. c.* (1899). Type: Angola, *Welwitsch* 7524 (BM, holotype).
Andropogon kimuingensis Vanderyst in Bull. Agric. Congo Belge 11: 144 (1920), *in synon.*

21a. var. **filipendula**

Caespitose perennial; culms 60 cm. to 2 m. high, branched from the lower nodes, arising from short scaly rhizomes. *Leaf-sheaths* glabrous, or rarely the basal sheaths sparsely pubescent with stiff white hairs; ligule up to 1 mm. long; blades up to 30 cm. long and 4 mm. wide. *Spathate panicle* virgate, 30–80 cm. long, the branching slender and graceful; spatheoles linear to almost filiform, 4·5–5·5 cm. long; peduncles very fine and flexuous, as long as or a little shorter

than the spatheole, with or without spreading white hairs towards the tip.
Racemes not deflexed, 10–12 mm. long, 2-awned per pair, delicate, yellowish-
green often tinged with violet; raceme-bases very unequal, the upper (4–)4·5–
8(–10) mm. long, slender, glabrous (rarely pubescent or with a few scattered
hairs). *Homogamous pairs* 1 at the base of the lower end 2 at the base of the upper
raceme; spikelets linear-lanceolate, 5–7 mm. long, glabrous. *Sessile spikelets*
linear-oblong, 5·5–7 mm. long, glabrous; callus 1·8–3 mm. long, pungent;
lower glume flat on the back or with the inner nerves more or less raised towards the tip and with an indistinct median hollow towards the base; awn
3–5·5 cm. long, the column hirtellous with fulvous hairs 0·7–1·2 mm. long.
Pedicelled spikelets linear-lanceolate, 5–6 mm. long, glabrous, terminating in a
bristle 1–5 mm. long; pedicel tooth very short, triangular.

MINOR VARIATIONS. A few specimens may have 3 or 4 awns per raceme-pair,
but they are otherwise similar to the normal form. The spikelets may be infected
by the smut *Sphacelotheca barcinonensis* Riofrio.

HABITAT. A common, often dominant, savanna species occurring on a wide
range of soil types, often forming small spreading tufts between the larger
species.

DISTRIBUTION. Throughout tropical Africa but mainly in the east; one specimen only from Asia (Map 7, p. 91).

TIMOR. Noiltoko (May), *Monod de Froideville* 1560.

GUINEA. Madina Tossekré (Oct.), *Adam* 12543, 12545; Fetoré (Oct.),
Adames 398; Mali (Dec.), *Schnell* 2358.

NIGERIA. Jos Plateau (Oct.), *Lely* P807; Jos town (Nov.), *Clayton* 1448;
Gindiri (Oct.), *Hepper* 1135; Panshanu Pass (Aug.), *Lawlor & Hall* 456; Biu
(Oct.), *de Leeuw* 1308.

CAMEROUN. Jakiri (June), *Brunt* 766; Mokolo (Oct.), *Letouzey* 7288 bis;
Ngaundéré to Poli (Oct.), *Koechlin* 7262; Meiganga to Ngaundéré (Oct.),
Koechlin 7209, 7221; Meiganga (Oct.), *Koechlin* 7144.

CENTRAL AFRICAN REPUBLIC. Bocaranga (Sept.), *Koechlin* 6155; Bouar
(Sept.), *Koechlin* 6245, 6268; Ippy (Oct.), *Tisserant* 1669; Bambari (Dec.),
Descoings 11802.

GABON. Booué (Aug.), *J. N. Davies* 317.

REPUBLIC OF CONGO. Massa (Nov.), *Sitha* 79; Mbé (Jan.), *Koechlin* 1701.

CONGO REPUBLIC. Parc Nat. Garamba (Oct.), *de Saeger* 1491; Kawa (Aug.),
H. B. Johnston 1121; Kivu, Kiteke (May), *Léonard* 4240; Kwango, Makayabo
(June), *Vanderyst* 3767; Kisantu, Kapondo (Apr.), *Callens* 1406.

RWANDA. Kibugabuga (Nov.), *Troupin* 9060; Mutara (Mar.), *Troupin*
6683; Nyanza (Jan.), *Shantz & Turner* 4220; Kibungu (Jan.), *Troupin* 5784.

SUDAN. Jebel Marra, Murtagello (June), *Blair* 70; Jebel Marra, Nyertete
(July), *Blair* 95; Yei to Kajo Kaji (Dec.), *M. N. Harrison* 510; Nimule (July),
Shantz 900; Kapoeta (Aug.), *Peers* KO22.

ETHIOPIA. L. Tana (Sept.), *Ferguson* 56; Jimma (June), *R. B. Stewart* 72.

UGANDA. West Nile, Mt. Otze (June), *A. S. Thomas* 1970; Karamoja,
Napyenya (Sept.), *Dyson-Hudson* 112; Tororo (Mar.), *Michelmore* 1195; Jinja
(June), *G. H. S. Wood*; Masaka (Oct.), *Maitland* 751.

KENYA. Hoey's Bridge (Aug.), *Smith & Paulo* 910; Meru (July), *Bogdan*
3752; Kisumu (Feb.), *Dummer* 1798; Machakos (Jan.), *D. C. Edwards* 1636;
Mombasa, *Turnbull* 6.

TANZANIA. Bukoba, Kamachumu (Mar.), *Haarer* 2489; L. Manyara, Bagayo R. (Mar.), *Greenway & Kanuri* 11383; W. Usambaras, Mombo to Soni (June), *Drummond & Hemsley* 3001; L. Kwela, *Bullock* 2958; Kondoa Irangi, Hundwi (Mar.), *B. D. Burtt* 1531; Songea (Apr.), *Milne-Redhead & Taylor* 9389; Zanzibar, *Boivin*.

MOZAMBIQUE. Mutuali (Apr.), *Balsinhas & Marrime* 451; Angónia (May), *Mendonça* 4220; Milange (Oct.), *Torre* 3380; Macequece (Mar.), *Mendonça* 1172; Bela Vista to Salamanga—Maputo (Nov.), *Torre* 2129.

MALAWI. Zombwe (Jan.), *G. Jackson* 367; Mzimba, *S. G. Wilson* 20; Lilongwe (June), *G. Jackson* 462; Zomba, *Cormack* 433; Shire highlands, *Adamson* 102.

ZAMBIA. Mporokoso, Mweru-wa-Ntipa (Apr.), *Phipps & Vesey-FitzGerald* 3260; Fort Jameson (Mar.), *Verboom* 542; Monze to Magoye (Feb.), *F. White* 7229; Choma (Apr.), *Astle* 2992; Livingstone (Feb.), *Mitchell* 24/83.

RHODESIA. Shangani, Gwampa F.R. (Feb.), *Goldsmith* 20/56; Salisbury, *Craster* 74; Rusapi (Feb.), *Norlindh & Weimarck* 5103; Matopos, *R. M. Davies* 78; Melsetter (June), *Crook* 477.

ANGOLA. Jalaquizanga (June), *Gossweiler* 8420; Malange plateau (Apr.), *Dawe* 388; Henrique de Carvalho (Apr.), *Gossweiler* 11539; Munongue (Apr.), *Gossweiler* 3119; R. Quiriri, *Gossweiler* 3719.

MADAGASCAR. Andriba (Apr.), *Bosser* 8002.

SOUTH WEST AFRICA. Linyanti (Dec.), *Killick & Leistner* 3175.

SWAZILAND. Skurvekop (Dec.), *Pole-Evans* 3386; Bremersdorp (Apr.), *Murdock* 148.

SOUTH AFRICA. Transvaal: Sibasa, Tshakhuma (May), *van Warmelo* 5159/16; Sibasa, Lukandwane (Feb.), *Codd* 6880; Letaba (Jan.), *Scheepers* 1198; Nelspruit (Feb.), *de Winter & Codd* 455, *Liebenberg* 2489. Natal: Without locality, *Krauss* 28; St. Lucia bay (July), *Schweickerdt* 1375; Hlabisa (May), *C. J. Ward* 2320; Richard's bay (Dec.), *Halse* 27; Stanger (Apr.), *Pentz & Acocks* 10351. Cape Province: Mafeking (Oct.), *Shantz* 237.

21b. var. **pilosa** (*Hochst.*) *Stapf* in Prain, Fl. Trop. Afr. 9: 324 (1918).

Andropogon filipendulus var. *pilosus* Hochst. in Flora 29: 115 (1846); Hack. in DC., Monogr. Phan. 6: 635 (1889). Type: South Africa, *Krauss* 164 (K, isotype).
Anthistiria fasciculata Thw., Enum. Pl. Zeyl.: 366 (1864). Type: Ceylon, *Thwaites* 940 (K, holotype).
Andropogon lachnatherus Benth., Fl. Austral. 7: 534 (1878). Type: Australia, *Mueller* (K, lectotype).
A. finitimus var. *rectirameus* Hack. in Bol. Soc. Brot. 3: 137 (1885), *nom. nud.* Type: Angola, *F. Newton*.
A. filipendulus var. *lachnatherus* (Benth.) Hack. in DC., Monogr. Phan. 6: 635 (1889).
A. filipendulus var. *thwaitesii* Hack. in DC., Monogr. Phan. 6: 635 (1889). Based on *Anthistiria fasciculata* Thw.
A. filipendulus f. *bispiculatus* Hack. in Philipp. Journ. Sci. 1 (Suppl.): 267 (1906). Type: Luzon, *Merrill* 4298 (K, isotype).
Hyparrhenia piovanii Chiov. in Atti R. Acad. Ital., Mem. Cl. Sci. Fis. 11: 63 (1950). Type: Ethiopia, *Piovano* 66 (FI, holotype).

Racemes 2–4-awned per pair; spikelets pubescent to villous with white hairs. Otherwise similar to var. *filipendula*.

CHROMOSOME NUMBER. South Africa, 2n = 40 (de Wet & Anderson, 1956).

HABITAT. As for var. *filipendula*.

DISTRIBUTION. Throughout tropical Africa, but rare west of the Cameroun Mts; also in eastern Asia from Ceylon to Australia.

CEYLON. Without locality, *Thwaites* 940.

CELEBES. Lombasang (May), *Bunnemeyer* 11671.

PHILIPPINES. Luzon: Central region, *Loher* 1804; Twin Peaks (May), *Elmer* 6392; Ambuklas to Daklan (Oct.–Nov.), *Merrill* 4398; Baguio (Mar.), *Elmer* 8533; Bued R. (Oct.–Nov.), *Merrill* 4298.

NEW GUINEA. Eastern highlands, Goroka (Nov.), *McKee* 1526; Eastern highlands, Miruma (June), *Hoogland & Pullen* 5378.

AUSTRALIA. Mareeba (June), *Blake* 9488; Mackay, Middle Percy Is. (Jan.), *Lazarides* 5681; Islands of Moreton Bay (Aug.), *F. Mueller*; Tambourine Mt. (Mar.), *Domin*; Dungog (May), *Story* 7348.

NIGERIA. Naraguta (July), *Lely* P433; Okene (Nov.), *Clayton* 581.

CONGO REPUBLIC. Boma (Nov.), *Vanderyst* 27100; Matadi to Boma (May), *Bolema* 993; Gety (Aug.), *H. B. Johnston* 1064b; Katanda (Aug.), *Lebrun* 7515; Kalehe (Feb.), *Léonard* 3029.

RWANDA. Bugesera (Feb.), *Troupin* 6523; Mutara (Dec.), *Troupin* 9578; Mimuli (Apr., May), *Troupin* 7188, 7598; Lugadsi (Jan.), *Lebrun* 9527.

BURUNDI. Kiremba to Bweru (Feb.), *Van der Ben* 2458; Karuzi (Jan., Apr.), *Van der Ben* 1816, 2045; Karuzi to Nyabikere (Feb.), *Van der Ben* 1910; Burarana (Jan.), *Van der Ben* 1760.

SUDAN. Jebel Marra, Gollol (June, July, Sept.), *Blair* 45, 103, *Wickens* 2543; Jebel Marra, Beldong (Dec.), *Pettet* J30; Torit (Sept.), *Beshir* 442.

ETHIOPIA. L. Tana, *Grabham*; Harar to Jijiga (July), *Burger & Getahun* 357; Galla Sidama, *Piovano* 66.

UGANDA. Napak (May), *A. S. Thomas* 3642; Old Entebbe (Feb.), *Harker* 251; Katwe (June), *A. S. Thomas* 4155; Masaka to Mbarara (Jan.), *Snowden* 1336; Kambuga (Apr.), *A. S. Thomas* 3799.

KENYA. Kitale (Sept.), *Bogdan* 1953; Eldoret (Sept.), *Hitchcock* 25016; Kayamakago (Aug.), *Greenway* 7867; Thika (Jan.), *Bogdan* 1492; Makindu (Feb.), *Linton* 105.

TANZANIA. Tabora (Apr.), *Greenway* 10128; Bumbuli (May), *Drummond & Hemsley* 2510; Iringa, *Emson* 415; Rukwa, Ilemba Gap (Mar.), *McCallum Webster* T64; Songea distr., Gumbiro (May), *Milne-Redhead & Taylor* 10032.

MOZAMBIQUE. Vila Coutinho (May), *Mendonça* 4158; Angónia (May), *Mendonça* 4184; Nhacoongo (Mar.), *Macedo & Balsinhas* 1099; Moamba (Feb.), *Torre* 7381; Namaacha Mts. (Jan.), *Torre* 7086.

MALAWI. Mzimba (Mar.), *G. Jackson* 440; Lilongwe (Apr.), *G. Jackson* 461; Dedza (Mar.), *G. Jackson* 427; Ncheu (Mar.), *G. Jackson* 417; Zomba (May), *Wiehe* N/107.

ZAMBIA. Kawimbe (Mar.), *McCallum Webster* A195; Mwinilunga distr., Matonchi R. (Nov.), *Milne-Redhead* 3123; Mufulira (May), *Eyles* 8380; Broken Hill (May), *F. A. Rogers* 7701; Mazabuka (Jan.), *Trapnell* 859.

RHODESIA. Salisbury (Jan.), *Reynolds* 2454; Inyanga, Cheshire (Jan.), *Norlindh & Weimarck* 4369; Melsetter (Feb.), *Crook* 508; Ft. Victoria (Dec.), *D. A. Robinson* 160; Shangani (Mar.), *Goldsmith* 39/56.

BOTSWANA. Lobatsi (Apr.), *de Beer* Mog. 5.

ANGOLA. Henrique de Carvalho (Apr.), *Gossweiler* 11449; Nova Lisboa (May), *Gossweiler* 10747; Munongue (Apr.), *Gossweiler* 4150; Humpata (May), *Gossweiler* 10757; Huilla (May), *Pearson* 2718.

SOUTH WEST AFRICA. Andara (Feb.), *de Winter & Marais* 4784; Otavi to Grootfontein (Mar.), *de Winter* 2863.

SWAZILAND. Stegi (Feb.), *Compton* 29946, 26562; Hlatikula (Jan.), *Compton* 26381, 26427.

SOUTH AFRICA. Transvaal: Pilansberg, Buffelspan (Dec.), *Pole-Evans* 646; Pienaar's River (Mar.), *Pole-Evans* 704; Nylstrom (Dec.), *Pole-Evans* 676; Duivelskloof (Feb.), *Scheepers* 1203; Pilgrim's Rest (Feb.), *Wager* C58. Natal: Berea (Mar.), *J. M. Wood* 5933; Umpumulo (Feb.), *Buchanan* 223; Pietermaritzburg, Amatuba forest (Feb.), *Killick* 332; Durban, *Krauss* 164.

H. filipendula is recognized in its typical form most readily by its delicate and graceful appearance, to which the copious branching, slender ascending branches, filiform flexuous peduncles, and numerous small racemes with a long upper raceme-base and only 2 awns per pair all contribute. It may be confused with *H. poecilotricha* which has rufous spikelet hairs, and *H. finitima* which has a short upper raceme-base and at most 1 homogamous pair on the upper raceme.

Two varieties are commonly recognized according to whether the spikelets are glabrous or pilose, but the value of this distinction has been called into question by the observation that the two forms grow together and intergrade. There is however another difference, for whereas var. *filipendula* is a neatly circumscribed taxon, var. *pilosa* seems to merge completely with *H. hirta*. We can arrive at a circumscription of the species by studying first the typical 2-awned form of var. *filipendula*. This is shown in figure 23 (p. 100) where it is plotted against the typical form of *H. hirta*. It can be seen that, apart from the indumentum, there are a number of distinguishing features. The most important are the long shaggily hairy awns of *H. filipendula*, and its long callus; the latter is difficult to measure accurately owing to uncertainty as to where callus and glume join, but approximation is no disadvantage in view of the amount of variation within the panicle. The most distinctive feature of *H. filipendula*, the long upper raceme-base, turns out to be surprisingly uncritical in its intermediate range. The awned pedicelled spikelet of *H. filipendula* is a useful, though unreliable, additional character. It will be noted that the very few 3- or 4-awned specimens of var. *filipendula* fit comfortably into the main population.

Turning now to var. *pilosa* (Fig. 24, p. 101) it can be seen that the 2- and 3-awned specimens have parameters very similar to var. *filipendula*, with which they are obviously conspecific. The 4-awned specimens are much more variable, being distributed throughout the 2-awned range, and extending to a point well outside it; specimens with 5 or 6 awns lie wholly outside it. Inspection shows that the limit of the 2-awned range corresponds very closely with a transition from shaggily hairy awns to those with short appressed hairs. It is proposed that the definition of *H. filipendula* should rest upon this character, the 4-awned specimens being partitioned accordingly between *H. filipendula* var. *pilosa* and a new species *H. anamesa*.

Var. *pilosa* is distinguished not only by its spikelet indumentum, but also by the higher frequency of 3- and 4-awned specimens. It is here upheld because,

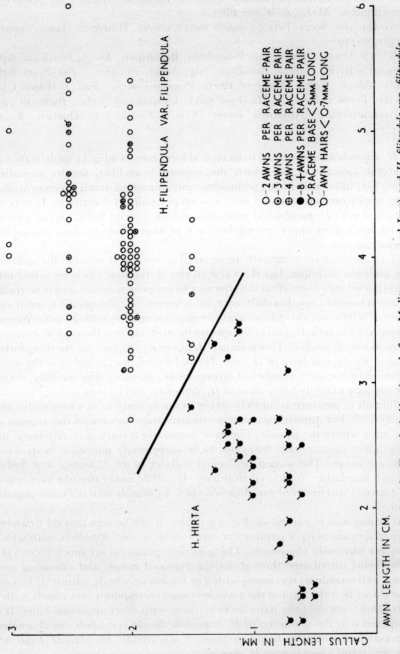

Fig. 23. Scatter diagram for *Hyparrhenia hirta* (sample from Mediterranean and Iraq) and *H. filipendula* var. *filipendula*.

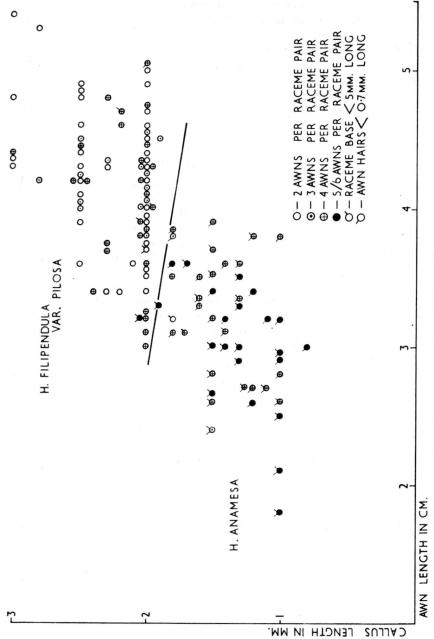

FIG. 24. Scatter diagram for: *Hyparrhenia filipendula* var. *pilosa* and *H. anamesa*. All the specimens plotted have two homogamous pairs at the base of the upper raceme.

although imposing a somewhat artificial boundary within the species, it provides a useful indication of the introgression of *hirta*-like characters. The isotype of var. *pilosa* at Kew is a glabrous specimen; conversely Hochstetter
remarks that some specimens from the type collection of var. *filipendula* have
hirsute glumes. It seems clear that both numbers are mixed gatherings, and
should be interpreted according to Hochstetter's circumscription.

22. **Hyparrhenia andongensis** *(Rendle) Stapf* in Prain, Fl. Trop. Afr. 9: 373
(1918).

Cymbopogon andongensis Rendle, Cat. Afr. Pl. Welw. 2: 159 (1899). Type:
Angola, *Welwitsch* 2728 (K, isotype).

Andropogon andongensis (Rendle) K. Schum. in Just, Bot. Jahresber. 27, 1: 454
(1901).

Perennial; culms wiry, 80 cm. high, arising from a creeping branched underground rhizome. *Leaf-sheaths* loosely hirsute; ligule up to 3 mm. long; blades
up to 22 cm. long and 2–4 mm. wide, puberulous to scaberulous above with
tubercle-based hairs towards the base, glabrous beneath. *Spathate panicle* about
30 cm. long, loose, rather scanty; spatheoles narrowly linear, 3–4 cm. long, dull
reddish; peduncles about ⅔ as long as the spatheole, glabrous. *Racemes* not
deflexed, 2·5 cm. long, 4–5-awned per pair, laterally exserted; raceme-bases
appendaged, unequal, the upper 4–5 mm. long, glabrous, filiform; appendage
scarious, narrowly oblong, 1 mm. long. *Homogamous pairs* 1 at the base of the
lower raceme; spikelets narrowly lanceolate, 7 mm. long, glabrous. *Sessile
spikelets* narrowly lanceolate, 7 mm. long; callus narrowly oblong, 0·7 mm. long,
obtuse; lower glume glabrous, 9-nerved, the nerves raised to form prominent
longitudinal ribs; awn 2 cm. long, the column puberulous with hairs 0·1 mm.
long. *Pedicelled spikelets* narrowly lanceolate, 8 mm. long, glabrous, purplish,
acuminate; pedicel tooth obscure.

DISTRIBUTION. Confined to Angola (Map 7, p. 91).

ANGOLA. Pungo Andongo, *Welwitsch* 2728.

H. andongensis can perhaps be compared with *H. griffithii*, for it bears as much
resemblance to this as to any other species in the section. It would be truer to
regard it as set apart from the rest of the section by its appendaged raceme-base
and longitudinally ridged lower glume. Both these characters are found in sect.
Hyparrhenia, but the long glabrous filiform raceme-base of *H. andongensis*
scarcely qualifies it for this section. Stapf (1918) placed the species in sect.
Apogonia adjacent to *H. gossweileri*. There is indeed some similarity, but whereas the raceme-base of the latter is broadly flattened, that of *H. andongensis* is
decidedly terete. It seems that the species will be anomalous wherever it is
placed, but that its inclusion in sect. *Polydistachyophorum* at least has the merit of
conforming to the major key characters.

III. Sect. **Pogonopodia** *Stapf* in Prain, Fl. Trop. Afr. 9: 293 (1918); Pilger
in Engl., Nat. Pflanzenfam. 14e: 174 (1940). Lectotype species: *H. cymbaria*.

Hyparrhenia series *Cymbariae* Stapf, *l. c.*: 293 (1918). Type species: *H. cymbaria*.

Perennials (except *H. anthistirioïdes*); culms typically robust, but sometimes
slender or even rambling. *Spathate panicle* typically large and decompound, less
often scanty; spatheoles short, linear-lanceolate to narrowly ovate, brightly
coloured, usually enclosing the shortly peduncled raceme-pairs (peduncles

longer in *H. papillipes, H. pilgerana, H. cyanescens*). Racemes deflexed, though sometimes tardily so (especially *H. pilgerana*), 3–8(–25)-awned per pair, glabrous or hoary villous, never rufous or fulvous; raceme-bases subequal, short, mostly 1–1·5 mm. long, flattened, copiously bearded with stiff glassy white or yellowish bristles, the tip often furnished with a lobed scarious rim (a definite appendage in *H. papillipes*). *Homogamous pairs* 1 at the base of the lower raceme only; spikelets narrowly lanceolate, villous, or glabrous but then ciliolate on the margins. *Sessile spikelets* lanceolate to lanceolate oblong, 4–6 mm. long; callus short, square or broader than long to cuneate, broadly rounded to subacute; upper lemma lobes short, 0·2–0·5 mm. long; awn up to 4·5 cm. long, the column puberulous to pubescent with yellowish hairs 0·2–0·4 mm. long (except *H. anthistirioïdes*). *Pedicelled spikelets* narrowly lanceolate, muticous or with a short awn-point (aristulate in *H. rudis, H. schimperi, H. cyanescens*), without a callus; pedicel tooth obscure (except *H. papillipes*).

The section is distinguished by the short, flat, bearded raceme-bases, and by the single homogamous pair. It shares these characters with sect. *Hyparrhenia*, the latter being distinguished by its appendaged raceme-bases. Unfortunately this character is not as precise as it may seem, for the raceme-base in sect. *Pogonopodia* usually has a scarious rim which may assume the dimensions of a small appendage. There is a clearly discernible difference in size (except in the case of *H. papillipes*), but the transition from an oblique tip to a scarious appendage is very gradual, and there is so much latitude for subjective interpretation that measurements may be misleading. Sect. *Pogonopodia* is also characterized by many-awned racemes enclosed in broad, coloured spatheoles and by a short callus; as opposed to the few-awned racemes, narrow spatheoles and pungent callus of sect. *Hyparrhenia*. These characters leave no doubt that species such as *H. schimperi, H. dregeana* and *H. papillipes* should be included in sect. *Pogonopodia* despite the presence of a small appendage; indeed it is most difficult to separate them from other species in this section.

In this section the upper raceme-base is flat, up to 1·5 or rarely as much as 2 mm. long (up to 2·5 mm. in *H. papillipes*), and obscured by stiff bristles. It approaches closely to sect. *Polydistachyophorum*, but here the raceme-base is terete, never less than 1·5 mm. long and usually much longer, and the hairs, if any, are usually rather scanty. A comparison between analogous species in the two sections shows that none of them are coincident with regard to their other characters, and we may fairly conclude that the nature of the raceme-base provides a valid, though occasionally rather tenuous, distinction between the two sections.

This section displays, in an extreme form, some of the problems which make *Hyparrhenia* such a difficult genus to understand. At first sight a number of distinct species can apparently be recognized, but attempts at precise definition reveal that they overlap and intermingle to a disconcerting degree.

Undoubtedly part of the trouble lies in the need to use qualitative characters (such as callus shape and spikelet hairiness) which allow a good deal of latitude for personal judgement or, as in the case of basal parts, are inadequately represented in herbarium specimens; and quantitative characters (such as awn and spatheole length) which vary enormously in the same panicle, and are affected to an unknown degree by environment. A conscious attempt to select an average spatheole from a panicle for measurement, or the greatest number of awns per raceme-pair in the panicle, is but an imperfect solution when the species are so narrowly separated. Quite apart from these practical difficulties, it seems more than likely that the blurred boundaries between species are due to genetic complications, such as introgression or apomixy, whose effect can only be surmised at present.

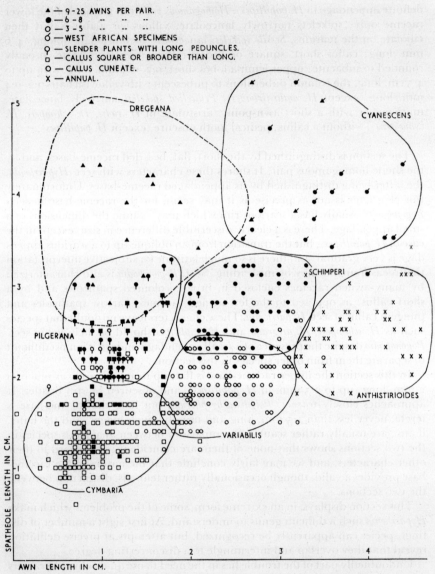

FIG. 25. Scatter diagram for species of *Hyparrhenia* sect. *Pogonopodia* with glabrous pedicelled spikelets.

Whatever the cause, we are faced with a situation in which exclusive characters and sharp discontinuities, are apparently absent. Instead there is a continuum broken into populations by peaks in the distribution curves of the quantitative characters, and by significant changes in the frequency of qualitative characters. It can be argued that such nebulous differences call for judicious lumping, an argument which in this case snowballs irresistibly until the section is reduced to a single aggregate species. Such a solution is by no means satisfactory owing to the very wide range of variation which it would embrace; nor can we ignore the tendency for the populations to be concentrated about certain noda to which many competent taxonomists have accorded the rank of species. The dilemma can readily be appreciated from figures 25 and 26. It seems to me that to sweep the pieces under the carpet of

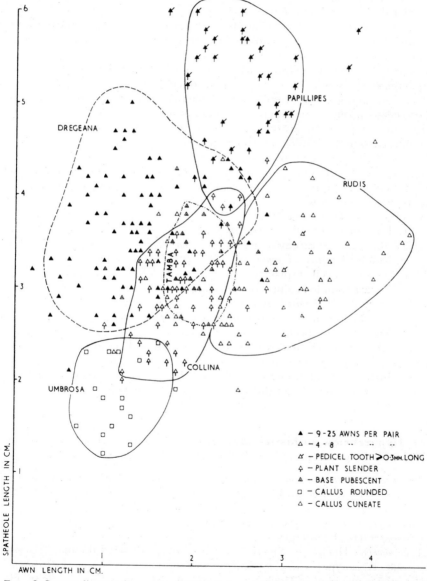

Fig. 26. Scatter diagram for species of *Hyparrhenia* sect. *Pogonopodia* with villous pedicelled spikelets.

an aggregate species would be a backward step, and that, despite the obvious difficulties, it is better to employ a classification which reflects the morphological diversity of the group. However the consequent difficulties should be appreciated:

a. In the absence of sharp breaks the boundary between species must necessarily be defined somewhat arbitrarily, and the identity of marginal specimens will be liable to the uncertainty of subjective decisions.

b. Diagnoses and keys, unless so widely drawn that they are useless for discrimination, will inevitably exclude a small proportion of specimens. These can only be assigned to the correct species by a consideration of overall similarities.

c. The chosen rank of species is obviously open to question. This cannot be cleared up until genetic mechanisms in the section are better understood, and it is pointless to debate it now.

The most obvious division within the section is that between hairy and apparently glabrous racemes, but is this distinction of taxonomic significance? To test this proposition the specimens with hairy (defined as possessing villous pedicelled spikelets) and more or less glabrous racemes were worked out separately. Upon comparison *H. dregeana* was the only species to show exact counterparts in the two classifications and, with this exception, the distinction between hairy and glabrous pedicelled spikelets was taken to be of taxonomic significance. Needless to say there are some further difficulties which are discussed under the appropriate species. Certain groups of specimens may then be removed, as follows:

A. Villous pedicelled spikelets.
    1. Long pedicel tooth (*H. papillipes*)
    2. Many-awned raceme-pairs (*H. dregeana*)
    3. Rounded callus (*H. umbrosa*)
B. Glabrous pedicelled spikelets
    1. Square rounded callus (*H. cymbaria*)
    2. Annuals (*H. anthistirioïdes*)
    3. Rambling plants with long peduncle (*H. pilgerana*)
    4. West African specimens (*H. cyanescens*)

The populations indicated by these more distinctive characters then provide the nuclei about which we may seek to establish boundaries, and by analogy with which we may attempt to treat the remainder of the complex. An overall view of the classification of the section is provided by figures 25 (p.104) and 26 (p. 105), and by Table 7.

23. **Hyparrhenia anthistirioïdes** (*Hochst. ex A. Rich.*) *Anderss. ex Stapf* in Prain, Fl. Trop. Afr. 9: 331 (1918).

Andropogon anthistirioïdes Hochst. ex A. Rich., Tent. Fl. Abyss. 2: 463 (1851); Hack. in DC., Monogr. Phan. 6: 630 (1889). Type: Ethiopia, *Schimper* 1822 (K, isotype).

Anthistiria pseudocymbaria Steud., Syn. Pl. Glum. 1: 399 (1854). Type: Ethiopia, *Schimper* 923 (K, isotype).

A. quinqueplex Hochst. ex Steud., *l. c.*: 400 (1854). Type: Ethiopia, *Schimper* 1098 (K, isotype).

Hyparrhenia quinqueplex (Hochst. ex Steud.) Anderss. in Schweinf., Beitr Fl. Aethiop.: 300 (1867), *comb. inval.*

TABLE 7. Characters of sect. *Pogonopodia*

	Culms slender or annual	Basal sheaths pubescent	Pedicelled spikelets villous	Spatheole length cm.	Peduncle length mm.	Awn length cm.	Awns/ pair	Callus square	Pedicelled spikelet usually awned	Appendage usually noticeable
H. anthistirioïdes	A	−	−	1·8–3·2	5–10	3·2–4·5	3–4(–5)	−	+	−
H. cymbaria		±	−	0·8–1·8(–2·1)	3–8	0–1·6(–2)	3–5(–6)	+	−	−
H. variabilis	S	+	−	1·4–2·4	3–9	1·8–3·2	3–5	−	+	−
H. pilgerana		−	−	2–3	9–30	0·7–1·7	(4–)6–7	±	−	−
H. formosa		−	−	1·8–2·6	2–10	0·8–1·8	6–8	±	−	−
H. schimperi		±	−(+)	2·2–3·2	10–15	2–3·3	6–8	−	+	+
H. cyanescens	S	−	−	3·5–5	10–50	2·8–3·4	6–11	+	−	−
H. papillipes		±	+	4–7	30–80	2–3	9–19	−	+	‡
H. dregeana		+	±	2·5–5	15–50	0–2(–2·8)	10–25	−	−	+
H. tamba		+	+	2·6–4	20–30	1·6–2·5	5–8	−	−	+
H. umbrosa	S	−	+	1·2–2·3	3–13	0·7–1·3	4–6	±	−	−
H. rudis		−	+	2·5–4	10–20	2·2–4	4–7	−	+	−
H. collina		−	+	2–4	10–25	1·5–2·5	4–7	−	−	−

Sorghum anthistirioïdes (Hochst. ex A. Rich.) Kuntze, Rev. Gen. Pl. 2: 791 (1891).

Hyparrhenia pseudocymbaria (Steud.) Stapf in Prain, Fl. Trop. Afr. 9: 329 (1918).

Annual; culms 30–150 cm. high, or shorter on the poorest soils, geniculate or ascending, sometimes with stilt roots from the lowest nodes. *Leaf-sheaths* glabrous; ligule about 1 mm. long; blades up to about 30 cm. long and 10 mm. wide, glabrous or very sparsely hairy, somewhat flaccid. *Spathate panicle* ample and 20–30 cm. long, or sometimes reduced to a few pairs of racemes, the coloured spathes and spatheoles contrasting with the dark awns; spatheoles lanceolate in profile, 1·8–3·2 cm. long, thinly scarious, often gaily streaked in shades of green, yellow and orange-brown; peduncles 5–10 mm. long, $\frac{1}{4}$–$\frac{1}{2}$ as long as the spatheole, bearded above. *Racemes* 1·1–1·3 cm. long, 3–4(–5)-awned per pair, exserted laterally; raceme-bases subequal, the upper about 1 mm. long, with or without a scarious lobe up to 0·2 mm. long. *Homogamous spikelets* 8–11 mm. long, glabrous except for the ciliate margins. *Sessile spikelets* 5–6 mm. long, glabrous or sparsely pubescent; callus cuneate, subacute, 0·5—1 mm. long; awn 3·2–4·5 cm. long, the column usually dark brown to black and hirtellous with hairs 0·8–1 mm. long. *Pedicelled spikelets* 5–6 mm. long, glabrous except for the ciliolate margins, finely aristulate with an awnlet 3–6 mm. long, sometimes with a perceptible callus up to 0·3 mm. long or even more.

ILLUSTRATION. Pilger in Engl. & Prantl, Nat. Pflanzenfam. 14e: 175, fig. 93a (1940).

MINOR VARIATIONS. *Pielou* 174 and *Burnett* 49/110 are two unusually robust specimens (about 2 m. high) from Milepa, Tanzania, with awns 3·4–4·5 cm. long. The awns are too long for *H. variabilis*, and it seems that the size range of *H. anthistirioïdes* should be extended to include these plants with stout culms. Unhappily these specimens lack the basal parts needed to confirm their identity.

Lynes 212 from Iringa, Tanzania, is a similar robust plant without a base. It is remarkable in having a pair of homogamous spikelets at the base of both racemes.

HABITAT. Rocky hillsides, volcanic tuffs, riverine alluvium and old cultivation. Probably a pioneer species of bare soils.

DISTRIBUTION. North-east tropical Africa from Jebel Marra to Somalia, and in Tanzania; but apparently absent from the intervening regions of Kenya & Uganda (Map 8).

SUDAN. Jebel Marra (Jan., Nov.), *Wickens* 1059, *Lynes* 180, *J. K. Jackson* 3873; Roseires, *Broun* 39; Gedaref (Oct.), *Beshir* 35.

ERITREA. Keren, *Colville* 85, *M. N. Harrison* 1305; Hawasien, *Baldrati*; Sarae (Oct.), *Pappi* 140; Adi Mendad (Sept.), *Pappi* 8519.

ETHIOPIA. Adoa (Nov.), *Schimper* 1098; Gennia (Dec.), *Schimper* 923; Schire highlands (Oct.), *Schimper* 1822; Gondar (Oct.), *Chiovenda* 2367; Addis Ababa to Moggio (Nov.), *Meyer* 7439; Harar (Oct.), *Burger* 1085; Nadda to Soddu (Oct.), *Mooney* 6101.

SOMALI REPUBLIC. Hargeisa (Oct.), *Hemming* 2250.

TANZANIA. Musoma, Beacon (June), *Greenway & Turner* 10363; Mbulu, Jebogo (Apr.), *B. D. Burtt* 2570; Mpwapwa (May), *van Rensberg* 447; Mbeya (May), *Milne-Redhead & Taylor* 10066; Songea, Litenga hill (Apr.), *Milne-Redhead & Taylor* 9687.

MALAWI. Salima to Lilongwe (Apr.), *Wiehe* N485.
ZAMBIA. Abercorn, Kawimbe (Mar.), *McCallum Webster* A236.

H. anthistirioïdes is readily distinguished from other members of the section by its annual habit. It should be noted, however, that in some of the perennial species, notably *H. cymbaria*, the connection between the stilted culm and perennating rhizome is often exceedingly tenuous; it is easily broken by the collector, and then overlooked on the herbarium sheet. Usually the species can also be distinguished by its long spatheoles, awns and homogamous spikelets, but these tend to intergrade with *H. variabilis*. Other more nebulous, but useful, characters are the small stature, gay striped spatheoles, and dark awns.

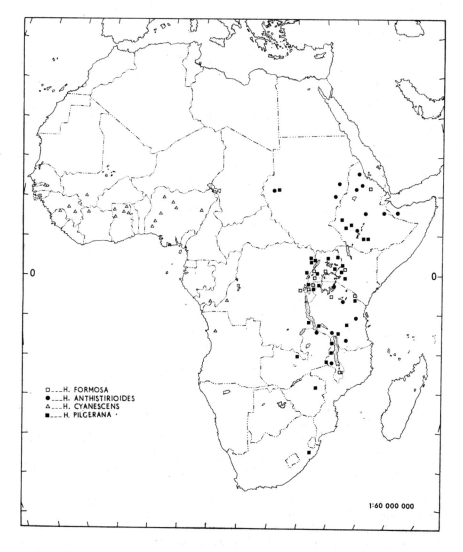

MAP. 8. Distribution of *Hyparrhenia formosa*, *H. anthistirioïdes*, *H. cyanescens* and *H. pilgerana*.

8

A distinction has been made between the smallest plants (*H. anthistirioïdes*) and the remainder (*H. pseudocymbaria*), but there is no difference in floral characters and I have no hesitation in uniting the two species. Mr. G. E. Wickens tells me that in the Sudan the tiniest plants are found on very shallow soils, their stature increasing on deeper and moister soils and occasionally reaching a height of 2 m. or more. In the light of these observations it seems likely that the aberrant specimens from Tanzania are simply robust plants from a particularly favourable site.

Anthistiria quinqueplex Hochst. ex Steud. was based on '*Andropogon cymbarius* var. *Cymbopogon quinqueplex*' Hochst. published, with diagnosis, on the herbarium label of *Schimper* 1098 in 1842. The intended rank and combination being quite uncertain, Hochstetter's original name must be disregarded.

24. **Hyparrhenia cymbaria** (*L.*) *Stapf* in Prain, Fl. Trop. Afr. 9: 332 (1918).

Andropogon cymbarius L., Mant. Alt.: 303 (1771); Hack. in DC., Monogr. Phan. 6: 629 (1889); Stapf in Thistleton-Dyer, Fl. Cap. 7: 360 (1898). Type: said to be from India, *Koenig* (LINN, holotype).

Cymbopogon elegans Spreng., Pug. Pl. Nov. 2: 14 (1815), *nom. superfl.*, based on *Andropogon cymbarius* L.

Anthistiria cymbaria (L.) Roxb., Fl. Ind. ed. Carey & Wallich, 1: 255 (1820).

Andropogon intonsus Nees, Fl. Afr. Austr. 1: 114 (1841). Type: South Africa (Umzimkulu), *Drege* (K, isotype).

A. lepidus Nees, *l. c.*: 113 (1841); Hack. in DC., Monogr. Phan. 6: 624 (1889). Type: South Africa (Omtento), *Drege* (K, isotype).

Cymbopogon cymbarius (L.) T. Thoms. in Speke, Journ. Discov. Source Nile: 652 (1863).

Anthistiria latifolia Anderss. in Peters, Reise Mossamb. Bot. 2: 562 (1864). Type: Comoro Is., *Peters*.

Andropogon lepidus var. *intonsus* (Nees) Hack. in DC., Monogr. Phan. 6: 625 (1889).

Sorghum cymbarium (L.) Kuntze, Rev. Gen. Pl. 2: 791 (1891).

S. lepidum (Nees) Kuntze, *l. c.*: 792 (1891).

Andropogon cymbarius var. *lepidus* (Nees) Stapf in Thistleton-Dyer, Fl. Cap. 7: 361 (1898).

Cymbopogon lepidus (Nees) Chiov. in Monogr. Rapp. Colon., Roma 24: 66 (1912).

Robust perennial, in coarse tufts from a slender creeping rhizome clad in small cataphylls; culms 2–4 m. high, initially slender and rambling, subsequently erect, stout (up to 8 mm. in diameter), and sustained by stilt roots. *Leaf-sheaths* usually glabrous, though often ciliate on the margins, and sometimes pubescent at the base; ligule up to 1 mm. long; blades mostly up to 45 cm. long and 6–20 mm. wide, rigid to subflaccid, dull green, glabrous or hirsute at the base. *Spathate panicle* large, dense, much-branched, typically 20–40 cm. long; spatheoles boat-shaped, narrowly ovate in profile, 0·8–1·8(–2·1) cm. long and 3–4 mm. wide, glabrous, turning a bright russet red at maturity; peduncles short, 3–8 mm. long, $\frac{1}{3}$–$\frac{1}{2}$ as long as the spatheole, bearded above with white or yellowish hairs. *Racemes* 0·7–1·3 cm. long, 3–5(–6)-awned per pair, projecting laterally from the spatheole; raceme-bases subequal, very short, up to 0·5 mm. long, the tip with or without a scarious frill up to 0·2 mm. long. *Homogamous spikelets* 4–6(–7) mm. long, glabrous to puberulous, ciliate on the margins. *Sessile spikelets* 3·8–4·5 mm. long, glabrescent to sparsely and shortly pubescent, often becoming purplish; callus square or broader than long, 0·2–0·3 mm. long, broadly rounded at the base; awn 0·5–1·6(–2·0) mm. long, rarely almost suppressed; caryopsis cylindrical, 3 mm. long. *Pedicelled spikelets*, 4–5 mm. long, glabrous to puberulous, ciliate on margins, acuminate or sometimes with an awn-point up to 1·5 mm. long.

ILLUSTRATIONS. C. E. Hubbard, E. Afr. Pasture Pl. 2: 10, f. 32 (1927); Pilger in Engl. & Prantl, Nat. Pflanzenfam. 14e: 175, f. 93b (1940); Andrews, Fl. Pl. Sudan 3: 472, f. 118 (1956).

MINOR VARIATIONS. Although normally seen as a tall robust plant, the initial shoot from the rhizome is slender and rambling, subsequently turning upwards, thickening, and becoming sustained by large stilt roots so that the original connection with the rhizome is scarcely evident. Under unfavourable conditions it can apparently flower while still in the rambling stage, thus simulating the habit of *H. pilgerana*.

The racemes usually have 3–5 awns per pair, but some specimens have 6, and *Bogdan* 2162 from Gilgil in Kenya has 8.

CHROMOSOME NUMBER. South Africa, 2n = 20 (de Wet, 1960). W. Cameroun, 2n = 20 *Brunt* 1269!. Madagascar, 2n = 30 *Tateoka* 3505! (in litt.).

HABITAT. A common species in tall grass savanna, with a preference for the upland regions.

DISTRIBUTION. The eastern half of Africa from Natal to Eritrea, reaching the west coast in the Cameroun area and in northern Angola (Map 9, p. 112). Also occurs in Madagascar and the Comoro Islands.

NIGERIA. Mambila plateau, Maisamari (Jan.), *Hepper* 1645.

CAMEROUN. Wum (Nov.), *Brunt* 876; Bafut-Ngemba F.R. (Mar.), *Onochie* FHI 34845; Bamenda (Jan.), *Migeod* 315; Bambui (Aug.), *Brunt* 1269; Bertoua (Dec.), *Baldwin* 13941; Hossere Ngo (Sept.), *Letouzey* 5692; Gounte (Jan.), *Letouzey* 2793; Poli (Jan.), *Raynal* 13102.

CONGO REPUBLIC. Mahagi Port (Oct.), *H. B. Johnston* 1179; Mumo (Jan.), *Léonard* 2666; Lacs Mokoto (July), *Van der Ben* 656; Rutshuru (Nov.), *Van Ysaker* 67; Tompa (Apr.), *Dubois* 1335; Kasiki (June), *Symoens* 4500.

RWANDA. Mutara (May), *Troupin* 7595.

ERITREA. Mt. Savour, *Pappi* 79; Mt. Bizem (May), *Schweinfurth & Riva* 2027.

ETHIOPIA. Semien (Dec.), *Chiovenda* 3166; Gondar (Jan.), *Pichi-Sermolli* 118; Jimma (Oct.), *Mooney* 6030, *Jimma Agric. School* 110; Agheremariam (Jan., Nov.), *Mooney* 5438, *Gillet* 14476.

SUDAN. Jebel Marra, Beldong (Dec.), *Pettet* J39; Lado, *Speke & Grant*; Yei (May), *Andrews* A1156; Oketch (Feb.), *Chipp* 27; Imatong Mts. (Feb.), *H. B. Johnston* 1488.

UGANDA. Kaboko (Oct.), *Hazel* 685; Masindi (July), *Langdale-Brown* 332; Mihunga (Jan.), *Loveridge* 345; Gayaza (Mar.), *Michelmore* 1352; Kigezi (Aug.), *A. S. Thomas* 2498; Kampala (Nov.), *Maitland* 161; Karamoja, Mt. Morongola (Nov.), *A. S. Thomas* 3261; Mt. Elgon (Aug.), *Snowden* 1162.

KENYA. Kitale (July), *Bogdan* 3502; Kipkarren to Kakamega (Aug.), *Pole-Evans & Erens* 164R; Archer's Post (June), *Hancock* 182; Meru (June), *Gardner* 2610; Embu to Nairobi (May), *Shantz* 1061; Kisii (Mar.), *Bogdan* 2949; SW. Mau forest (Jan.), *Kerfoot* 2741; Narok, Migori bridge (June), *Glover, Gwynne, Samuel & Tucker* 1803.

TANZANIA. Bukoba, Kidwe (Oct.), *Haarer* 2199; Kwangowe (June), *Volkens* 341; Kondoa, Kikori (Aug.), *B. D. Burtt* 2597; Mpwapwa (Apr.), *B. D. Burtt* 3953; Kigoma, Sisaga (Aug.), *Newbould & Jefford* 1943; Ufipa, Mbesi forest, (Mar.), *McCallum Webster* T129; Iringa, Mufindi (Aug.), *Greenway* 3472; Rungwe (June), *Stolz* 1353; Songea, Kigonsera (Apr.), *Milne-Redhead & Taylor* 9590.

MOZAMBIQUE. Gúruè (Sept., Nov.), *Mendonça* 2128, 2214; Zobue (June), *Torre* 2867; Chimoio (Apr.), *Mendonça* 1546; Spungabera (June), *Torre* 4262.

MALAWI. Nkata Bay (June), *J. D. Chapman* 731; Dedza Mt. (Sept.), *Burtt-Davy* 1481; Zomba (June, Aug.), *Brass* 16275, *Cormack* 279, *G. Jackson* 85; Shiri highlands, *Buchanan* 37.

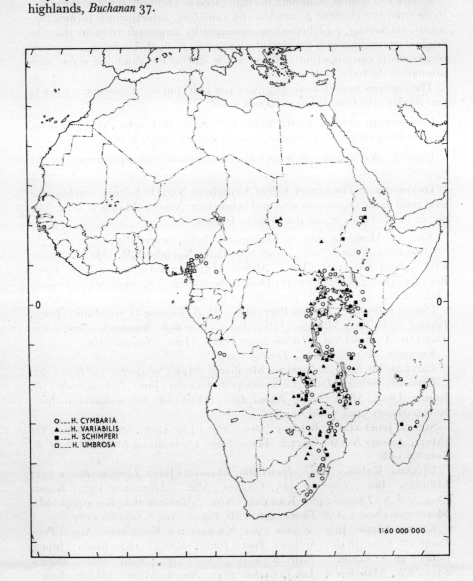

MAP 9. Distribution of *Hyparrhenia cymbaria, H. variabilis, H. schimperi* and *H. umbrosa*.

ZAMBIA. Mufulira (May), *Eyles* 8412; Ndola (Mar.), *A. C. R. Edwards* 16; Broken Hill (Aug.), *Trapnell* 1526; Chakwenga headwaters, 100–129 km. E. of Lusaka (Mar.), *E. A. Robinson* 6481.

RHODESIA. Inyanga, Mtarazi Falls (Apr.), *Phipps* 1163; Umtali, Masanga (Oct.), *D. A. Robinson* 351; Melsetter (Oct.), *Sturgeon & Panton* GH 30784; Mt. Selinda (July), *Michelmore* 292; Bulawayo, *Jeffreys* GH 41283.

ANGOLA. Without locality, *Welwitsch* 7300; Cuanza Norte, Zewza R. (Aug.), *Gossweiler* 8336; Cuanza Norte, Rianzando (Sept.), *Gossweiler* 8813.

MADAGASCAR. Without locality, *Lyall* 23; Central Madagascar, *Baron* 1098, 4900; Mevetanana, *Perrier de la Bâthie* 183; Tananarive (June), *Tateoka* 3505.

COMORO Is. *Bojer, Hildebrandt* 1717.

SWAZILAND. Pigg's Peak (Mar.), *Compton* 30622; Mbabane (May), *Pole-Evans* 3432; Usutu forest (Apr.), *Compton* 27775; Hlambanyati valley (May), *Compton* 25117.

SOUTH AFRICA. Transvaal: Devil's kloof (Apr.), *Pole-Evans*; Sibasa (May), *van Warmelo* 5157/4; Weltevreden (Mar.), *Scheepers* 190; Pilgrim's Rest, *F. A. Rogers* 18575; Sabie valley (Apr.), *Galpin* 13632. Natal: Laingsnek, *Rehmann* 6939; Eshowe (Apr.), *McClean* 983; Inanda, *J. M. Wood* 1304; Umpumulo (Nov.), *Buchanan* 228; Umzimkulu R. to Umkomanzi, *Drege.* Cape Province: Omtento to Umzimkulu, *Drège*; Port St. John (Apr.), *Galpin* 2860.

A conspicuous feature of *H. cymbaria* is its beautiful panicle of short, narrowly ovate, coloured spatheoles totally enclosing the shortly awned racemes. Unfortunately these characters are unsuitable for discrimination on anything but the most subjective basis, owing to the extent to which awn and spatheole size intergrade with adjacent taxa. However, the small-spatheoled specimens can also be segregated by the short rounded callus of the sessile spikelet and it is upon this character, admittedly still liable to some degree of subjective interpretation, that the species is based. It might be expected that lengths of spatheole and callus are causally correlated, both being possible consequences of short racemes, thus weakening the argument that the shape of the callus is subject to a discontinuity. However, on comparing *H. cymbaria* and *H. variabilis*, this objection is found to be scarcely tenable for callus shape is much more strongly correlated with awn length than with spatheole size.

The type of *Andropogon cymbarius* is a specimen of Koenig's 'ex India orientali'. However, the species has never been rediscovered in India, and it is likely that this specimen came from the Comoro Islands (Stapf, 1898: 362).

25. **Hyparrhenia variabilis** *Stapf* in Prain, Fl. Trop. Afr. 9: 334 (1918). Type: Zambia, *Macaulay* 62 (K, lectotype).

H. spectabilis Stapf, *l. c.*: 338 (1918). Type: Angola, *Welwitsch* 7247 (K, lectotype).

Cymbopogon acutispathaceus De Wild. in Bull. Jard. Bot. Brux. 6: 7 (1919). Type: Congo Republic, *Ringoet* 3 (BR, holotype).

Andropogon acutispathaceus De Wild., *l. c.* (1919), *in synon.*

Hyparrhenia acutispathacea (De Wild.) Robyns, Fl. Agrost. Congo Belge 1: 181 (1929).

H. iringensis Pilger in Not. Bot. Gart. Berlin 14: 101 (1938). Type: Tanzania, *Schlieben* 828.

Robust perennial, arising from a short underground rhizome clad in hard cataphylls; culms 1·5–3 m. high, supported by stilt roots, often initially decumbent. *Leaf-sheaths* glabrous, or sometimes pubescent at the base; ligule up to 2 mm. long; blades up to about 45 cm. long and 15 mm. wide, firm, glabrous or rarely hirsute at the base. *Spathate panicle* large, dense, much-branched, typically 20–40 cm. long; spatheoles boat-shaped, lanceolate in profile, 1·4–2·4 cm. long and 3–4 mm. wide, glabrous, russet red tinged with yellow and green at maturity; peduncles short, 3–9 mm. long, up to about ⅓ as long as the spatheole, bearded above with white hairs. *Racemes* 0·8–1·3 cm. long, 3–5-awned per pair, projecting laterally from the spatheole; raceme-bases subequal, short, the upper about 1 mm. long, with a scarious rim up to about 0·2 mm. long at the tip. *Homogamous spikelets* 7–9 mm. long, glabrous to puberulous, ciliate on the margins. *Sessile spikelets* 4–5 mm. long, glabrescent to sparsely and shortly pubescent; callus cuneate, 0·5–1 mm. long, narrowly obtuse to subacute at the base; awn 1·8–3·2 cm. long. *Pedicelled spikelets* 5–8 mm. long, glabrous to puberulous, ciliate on margins, with an awn-point 1–4 mm. long.

CHROMOSOME NUMBER. Uganda, 2n = 20 (Tateoka, 1965a)!. Zambia, 2n = 20 *Astle* 2989!.

HABITAT. Commonly a dominant constituent of tall grass savanna, occurring on a wide range of soil types.

DISTRIBUTION. Eastern Africa from Ethiopia to the Transvaal (Map 9, p.)112: also in Java.

JAVA. Without locality, *Buijsman* 52, 156.

CONGO REPUBLIC. Parc Nat. Garamba (Aug.), *Troupin* 1905; Kivu, Kabare *Gilon* 86; Bugarama (Mar.), *Germain* 6423; Marungu, Sakala (Apr.), *Dubois* 1148; Kipila (May), *Quarré* 1725; Shinsenda (May), *Ringeot* 3.

RWANDA. Parc Nat. Kagera (June), *Troupin* 3587.

SUDAN. Jebel Marra (May), *Wickens* 1567; Tonga, *Chipp* 5; Yei, Aloma plateau (Dec.), *M. N. Harrison* 613; Yei, Mt. Korobi (Dec.), *Myers* 7980; Torit, Katire, *J. K. Jackson* 469.

ETHIOPIA. Gondar (Jan.), *Pichi-Sermolli* 119.

UGANDA. West Nile, Paida (Aug.), *Chancellor* 211; Bunyoro, Kigorobya (May), *Tateoka* 3241; Hoima (Sept.), *H. B. Johnston* 1152; Ruwenzori, Nakitenga (Dec.), *Maitland*; Karamoja, Warr (Nov.), *A. S. Thomas* 3193; Tororo, Budama (July), *Maitland* 1179a.

KENYA. Mt. Elgon (Apr.), *Lugard* 613; Kitale (Dec.), *Bogdan* 3635; Makueni (Oct.), *Bogdan* 1401.

TANZANIA. Ngara, Kirushya (Nov.), *Tanner* 4520; Amani (Aug.), *Hitchcock* 24522; Uluguru Mts., Ngila (June), *Milne & Harris* 2; Mahali Mts., Utahya (Sept.), *Newbould & Jefford* 2382; Ufipa, L. Kwela (July), *Bullock* 2972; Rungwe to Mbosi (Mar.), *R. M. Davies* 361; Songea, Johannesbruck (Apr.), *Milne-Redhead & Taylor* 9689.

MOZAMBIQUE. Angónia (May), *Mendonça* 4188a; Massingire (May), *Torre* 5259; Chimanimani Mts. (Apr.), *Goodier* 996; Vila Pery (June), *Torre* 2793.

MALAWI. Fort Manning (Apr.), *G. Jackson* 771; Namweras (Apr.), *Lawrence* 385; Zomba, *Cormack* 277; Madzibango to Chikwawa (June), *Corbett* 415; Mlanje (June), *J. D. Chapman* 651.

ZAMBIA. Abercorn (Apr.), *Richards* 1537; Kasama (Apr.), *Vesey-FitzGerald* 1654; Fort Roseberry (May), *Astle* 3004; Solwezi (June), *Milne-Redhead* 491;

Mongu (Apr.), *Verboom* 1385; Mumbwa, *Macaulay* 62; Lusaka (May), *Trapnell* 2017; Choma, Mochipapa (Apr.), *Astle* 2989.

RHODESIA. Bindura (Mar.), *Rattray* 1492; Goromonzi (May), *Miller* 7295; Marandellas (Mar.), *Brain* 3229; Gokwe (Apr.), *Bingham* 630; Matopos, Konkora (Apr.), *Plowes* 1713; E. Victoria distr. (Feb.), *D. A. Robinson* 267.

ANGOLA. Cambondo, *Welwitsch* 7247; Benguella, *Gossweiler* 1714.

MADAGASCAR. Isalo (July), *Humbert* 5050; Tananarive (Dec.), *Bosser* 12474. COMORO Is. *Blackburn, Humblot* 147.

SOUTH AFRICA. Transvaal: Blauwberg (Dec.), *Smuts & Pole-Evans* 843; Pietersburg (Apr.), *Dyer* 3167; Potgietersrust (Mar.), *Galpin* 8887.

H. variabilis and *H. cymbaria* are obviously very closely related, for not only do the dimensions of their floral parts intergrade almost completely, but they share the same habit, distribution and chromosome number. The decision to maintain both species is guided chiefly by the fairly sharp difference in the shape of the callus, but also, let it be admitted, by an intuitive feeling that the distinctive little ovate spatheoles of typical *H. cymbaria* merit separation.

The type of *H. iringensis* has not been located, but its identity seems fairly clear from the description.

26. **Hyparrhenia pilgerana** (*'pilgeriana'*) C. E. Hubbard in Bull. Misc. Inf. Kew 1928: 39 (1928). Based on *Cymbopogon stolzii* Pilger.

Cymbopogon stolzii Pilger in Engler, Bot. Jahrb. 54: 286 (1917), *non Hyparrhenia stolzii* Stapf (1918). Type: Tanzania, *Stolz* 960 (K, isotype).
Hyparrhenia claessensii Robyns, Fl. Agrost. Congo Belge 1: 180 (1929), & in Bull. Jard. Bot. Brux. 8: 236 (1930). Type: Congo Republic, *Claessens* 1400 (K, isotype).

Perennial, forming lax tufts from a short rhizome clad in scaly white cataphylls; culms slender, 30–60 cm. high and 1–2 mm. diameter, weakly ascending or untidily straggling. *Leaf-sheaths* glabrous; ligule 1–2 mm. long; blades short, 5–10(–15) cm. long and 2–4 mm. wide, usually rather thin and light green, glabrous. *Spathate panicle* open and scanty, often with only 10–20 raceme-pairs but sometimes denser with more numerous pairs; spatheoles narrowly lanceolate, 2–3 cm. long, glabrous, reddish brown; peduncles 0·9–3 cm. long, from ½ as long to slightly exceeding the spatheole, pilose with yellow or sometimes white hairs towards the tip, exserted from the side of the spatheole with or without a sinuous hook. *Racemes* not, or tardily, deflexed, 1–1·5 cm. long, (4–)6–7-awned per pair; raceme-bases subequal and up to 0·5 mm. long, or rather unequal with the upper subterete and up to 1·5 mm. long, typically with a scarious rim up to 0·2 mm. long or rarely with a short appendage up to 0·5 mm. long. *Homogamous spikelets* 4–5·5 mm. long, glabrous except for the ciliolate margins. *Sessile spikelets* lanceolate, 4 mm. long, glabrous or white puberulous; callus oblong or sometimes square, 0·4–0·7 mm. long, obtusely rounded at the base; awn 0·7–1·7 mm. long. *Pedicelled spikelets* 4–4·5 mm. long, purple, glabrous, awnless or with an awn-point up to 1·5 mm. long.

ILLUSTRATION. Figure 27 (p. 116).

HABITAT. A species of seasonal swamps and forest margins; also of grassy hillsides and cultivated land, these perhaps being secondary habitats to which it extends when opportunity offers.

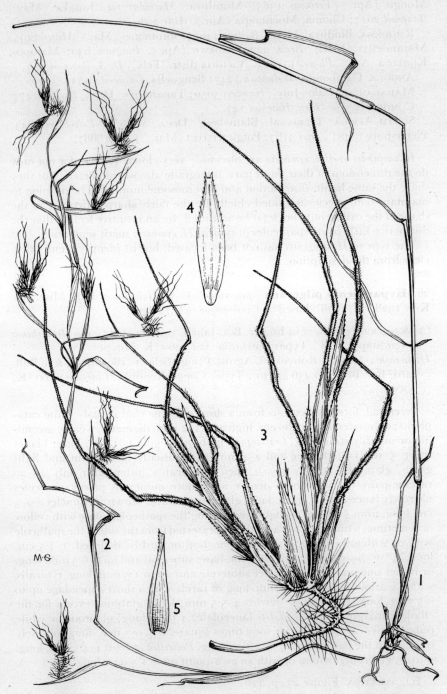

FIG. 27. *Hyparrhenia pilgerana*. **1,** base of plant showing habit of stem, ×⅔; **2,** inflorescence, ×⅔; **3,** raceme-pair, ×5. Sessile spikelet. **4,** lower glume, including callus, ×6; **5,** upper glume, ×6. **1,** from *Gillett* 14479, **2–5** from *Mooney* 8618.

DISTRIBUTION. Eastern Africa from Ethiopia to Malawi, with a few scattered specimens from further south (Map 8, p. 109).

CONGO REPUBLIC. Ituri, *Claessens* 1400; Nioka (Mar., Sept., Dec.), *Risopoulos, Froment* 241, *Taton* 1072; Mahagi to Djugu (Sept.), *Lebrun* 3832; Kasiki (June), *Symoens* 4422.

RWANDA. Nkuli (June), *Scaetta* 2253, *Humbert* 8622 bis.

SUDAN. Jebel Marra, Kurongo, *J. K. Jackson* 2628.

ETHIOPIA. Tulu Wallei (Feb.), *Mooney* 7759; Shabe (Nov.), *Mooney* 8618; Mt. Delo (Jan.), *Gillett* 55045; Agheremariam (Jan., Nov.), *Mooney* 5458, *Gillett* 14479; Shuka (Dec.), *Mooney* 8483.

UGANDA. Napak (May), *A. S. Thomas* 3648; Lango (June), *H. B. Johnston* 899; Nakishenyi (Feb.), *Liebenberg* 395; Tororo (Mar.), *Michelmore* 1199; Mbale (June), *Snowden* 1105.

KENYA. Kapenguria (Mar.), *Bogdan* 4513; Kitale (July), *Bogdan* 3501; Molo to Eldoret (Jan.), *Bogdan & Williams* 201; Mau forest (Jan.), *Bally* 4749; Ngerendei (Mar.), *Glover, Gwynne & Samuel* 155.

TANZANIA. Kitengule (Jan.), *Haarer* 2484; Kilimanjaro (Feb.), *Haarer* 1048; Mufindi, *R. M. Davies* B26; Mbesi forest (Mar.), *McCallum Webster* T133; Kyimbila (Nov.), *Stolz* 960.

MALAWI. Nyika (June), *G. Jackson* 529; Mzimba (Mar.), *G. Jackson* 433; Fort Manning (Jan.), *G. Jackson* 721.

ZAMBIA. Mufulira (May), *Eyles* 8393.

RHODESIA. Salisbury (Mar., Apr.), *Eyles* 4768, *Brain* 4242.

SOUTH AFRICA. Natal: Dundee, Waschbank R. (Feb.), *McClean* 101.

H. pilgerana is distinguished from *H. cymbaria* and *H. formosa* by its slender straggling culms and relatively long peduncles, which impart to it a distinctive facies. The effect is heightened by the colouring—purple spikelets framed in yellow hairs from the raceme-bases—which recalls part of sect. *Hyparrhenia*. The longer spatheoles also help to distinguish it from depauperate specimens of *H. cymbaria*. Other useful, but inconstant, characters are the tendency to a sinuously exserted peduncle and a subterete upper raceme-base.

It bears some resemblance to species in sect. *Polydistachyophorum*, notably *H. gazensis*, from which it is best distinguished by its shorter raceme-base and shorter awns.

27. Hyparrhenia formosa *Stapf* in Prain, Fl. Trop. Afr. 9: 340 (1918).
Type: Ethiopia, *Schimper* 1009 (K, holotype).

Andropogon formosus Hort.; Hack. in DC., Monogr. Phan. 6: 623. (1889), *in synon*; Stapf in Prain, Fl. Trop. Afr. 9: 341 (1918).

Robust perennial in coarse tufts sometimes reinforced by stilt roots; culms about 2 m. high. *Leaf-sheaths* glabrous; ligules 2 mm. long, truncate; leaf-blades up to 50 cm. long and 12 mm. wide, scaberulous. *Spathate panicle* large, dense, 30–40 cm. long; spatheoles lanceolate in profile, 1·8–2·6 cm. long, glabrous or rarely weakly pilose, russet red or tinged with purple and yellow; peduncles short, 2–10 mm. long, $\frac{1}{4}$–$\frac{1}{2}$ as long as the spatheole, bearded above. *Racemes* 1–1·5 cm. long, 6–8-awned per pair, exserted laterally; raceme-bases subequal, very short, up to 0·5 mm. long. *Homogamous spikelets* 4·5–5·5 mm. long, glabrous with ciliate margins. *Sessile spikelets* 4–4·5 mm. long, glabrous to shortly pilose; callus oblong to cuneate, rarely square, 0·5–1 mm. long, obtusely rounded; awn 0·8–1·8 cm. long. *Pedicelled spikelets* 4·5–6 mm. long, glabrous to puberulous, acuminate.

CHROMOSOME NUMBER. Kenya, 2n = 30 (Celarier, 1956)!.

HABITAT. Savanna grassland and streamsides.

DISTRIBUTION. A rather uncommon grass scattered throughout eastern Africa, but concentrated about Uganda (Map 8, p. 109).

CONGO REPUBLIC. Tshibinda, *Humbert* 7376 bis.
RWANDA. Mutara (Jan.), *Troupin* 5671; Astrida (Mar.), *Hendrickx* 7615.
ETHIOPIA. Tigré, *Schimper* 1009.
UGANDA. Ruwenzori, Bwamba Pass, *Osmaston* 3928; Ankole, Lwasamaire (Nov.), *H. B. Johnston* 1336a; Kampala (June), *H. B. Johnston.*
KENYA. Eldoret (June, Aug.), *Williams* 258, *Bogdan* 1910, 1911.
TANZANIA. Kilimanjaro, Maranga (Jan.), *Schlieben* 4633; Nyamwezi, Ugala Mbuga (Jan.), *Shabani.*
MALAWI. Zomba (Mar., Apr.), *Lawrence* 318, *Wiehe* N39.

H. formosa differs from *H. cymbaria* in its longer spatheoles and generally oblong to cuneate callus, while the number of awns per raceme-pair tends to be higher. These characters are similar to those in *H. pilgerana* but, with its stout culms and short peduncles, it cannot be accommodated in that species. *H. formosa* is, in truth, something of a hotch-potch, created by taking the *cymbaria*-like specimens with an oblong or cuneate callus and adding those specimens with a square callus whose spatheole is longer than the usual range of *H. cymbaria*. The distinction from *H. cymbaria* is obviously a fine one, and it may be that *H. formosa* is merely a group of specimens forming a peripheral extension to the range of *H. cymbaria*, a suspicion strengthened by the observation that the analogous species *H. umbrosa* covers the joint range of variation of the two glabrous species. However, the chromosome number (2n = 30) suggests that we may be attempting to segregate a higher ploidy level, for typical *H. cymbaria* has 2n = 20. Tateoka's count of 2n = 30 for *H. cymbaria* is not necessarily contradictory for it was made on a marginal specimen, and may simply indicate that the morphological boundary needs adjustment. On balance the evidence seems to favour the continued separation of these species.

The species, or probably a mixture of species originating from seed collected by Schimper in Ethiopia, was at one time grown in a number of botanical gardens and has been referred to in horticultural literature as '*Andropogon formosus*', but the name was never validly published.

28. **Hyparrhenia schimperi** (*Hochst. ex A. Rich.*) *Anderss. ex Stapf* in Prain, Fl. Trop. Afr. 9: 341 (1918).

Andropogon schimperi var. *longicuspis* Hochst. in sched., Schimp. Iter Abyss. 2: 1052 (1842), *nom. invalid.* [precedes publication of species]; Hochst. ex A. Rich., Tent. Fl. Abyss. 2: 467 (1851), *in synon.*
A. schimperi Hochst. ex A. Rich., Tent. Fl. Abyss. 2: 466 (1851); Hack. in DC., Monogr. Phan. 6: 623 (1889); Stapf in Thistleton-Dyer, Fl. Cap. 7: 357 (1898). Type: Ethiopia, *Schimper* 408 (K, isotype).
Sorghum schimperi (Hochst. ex A. Rich.) Kuntze, Rev. Gen. Pl. 2: 792 (1891).
Cymbopogon schimperi (Hochst. ex A. Rich.) Rendle, Cat. Afr. Pl. Welw. 2: 155 (1899).

Hyparrhenia viridescens Robyns, Fl. Agrost. Congo Belge 1: 182 (1929), & in
   Bull. Jard. Bot. Brux. 8: 238 (1930). Type: Congo Republic, *Quarré* 295
   (K, isotype).

Robust perennial, forming coarse tufts from a short underground rhizome;
culms 2–4 m. high, up to 8 mm. in diameter, often supported by stilt roots. *Leaf-
sheaths* glabrous, or rarely pubescent at the base; ligule up to 3 mm. long; blades
up to 60 cm. long and 2 cm. wide, glabrous or with a few hairs beneath.
Spathate panicle large, decompound, 30–60 cm. long; spatheoles narrowly
lanceolate in profile, 2·2–3·2 cm. long, glabrous, turning russet brown;
peduncles short, 10–15 mm. long, up to about half as long as the spatheole,
pilose towards the tip. *Racemes* 1·2–1·6 cm. long, 6–8-awned per pair (this being
the maximum number in a given panicle), internodes 2–2·5 mm. long, exserted
laterally; raceme-bases subequal, the upper 0·5–1·5 mm. long, the tip with a
scarious rim up to 0·6 mm. long. *Homogamous spikelets* 5–8 mm. long, glabrous
to sparsely pilose. *Sessile spikelets* 4–5 mm. long, glabrescent to sparsely pubescent; callus 0·5–0·8 mm. long, cuneate to acute; awn 2–3·3 cm. long. *Pedicelled
spikelets* 5–7 mm. long, glabrous to shortly and sparsely pilose, with, or sometimes without, an awn-point up to 6 mm. long.

CHROMOSOME NUMBER. South Africa, 2n = 40 (Krupko, 1955).

HABITAT. Savanna grasslands.

DISTRIBUTION. Eastern Africa from Ethiopia to Rhodesia; also in Madagascar (Map 9, p. 112).

CONGO REPUBLIC. Kambikila (Apr.), *Quarré* 295; Elizabethville (May,
July), *Symoens* 11004, 11005, *Burtt-Davy* 17807; Étoile (May), *Quarré* 395 bis.
   SUDAN. Jebel Marra (Dec.), *Pettet*.
   ETHIOPIA. Adoa (Nov.), *Schimper* 1052; Mt. Scholoda (Oct.), *Schimper* 408;
Gondar (Oct.), *Chiovenda* 2601; Asoso (Oct.), *Chiovenda* 2702; Garibaldi (Apr.),
Smeds 1145.
   UGANDA. Elgon, Mizembe (Aug.), *Snowden* 1168.
   KENYA. Nakuru (cult. at Kitale), *Muthambi* 5156; Kibwezi (Apr.), *Bally*
8091; Narok (July), *Glover & Samuel* 3051.
   TANZANIA. Moshi (June), *Haarer* 736; Kasulu, *Raunce* 5; Mpwapwa (June),
van Rensberg 471; Mufindi, *R. M. Davies* B28; L. Rukwa, Mwela (Apr.),
Michelmore 1022.
   MOZAMBIQUE. Without locality, *Hornby* 3347; Massingire (Apr.), *Torre*
5210.
   MALAWI. Mzimba, *S. G. Wilson* 18; Kasungu (Nov.), *G. Jackson* 683; Fort
Manning (Apr.), *G. Jackson* 772; Lilongwe (Mar., Apr.), *G. Jackson* 464, 1193.
   ZAMBIA. Abercorn, Ilembwe gorge (Apr.), *McCallum Webster* A359; Ndola
(Apr.), *G. Jackson* 1; Nangweshi (Aug.), *Codd* 7420; Mumbwa (Mar.), *van
Rensberg* 1690; Chilanga (Sept.), *C. Sandwith*; Lusaka (Nov.), *van Rensberg*
2624.
   RHODESIA. Kapanda (May), *Lovemore* 398; Kasulu (Oct.), *R. M. Davies*
1597a; Salisbury (Mar.), *Eyles* 1577; Zimbabwe (July), *Plowes* 1871; Selinda
(July), *Michelmore* 297.
   MADAGASCAR. Without locality, *Baron* 2304, 4973, 5068; Nossi Bé, *Boivin*
1983; Antsalova to Tsiandro (Oct.), *Bosser* 18171.
   SOUTH AFRICA. Natal: Umpumulo (Nov.), *Buchanan* 229.

Just as *H. cymbaria* grades into *H. variabilis* so, along a different axis, does *H. variabilis* grade into *H. schimperi* (Fig. 25, p. 104), all three species sharing the same geographical distribution. Indeed it is probable that the division between the two latter species, based on the number of awns per raceme-pair, is largely arbitrary, but nevertheless it serves to segregate a fairly compact cluster of specimens. The justification for this procedure lies in Krupko's chromosome count, from which it seems likely that we are attempting to separate tetraploid plants from the diploid *H. variabilis*.

H. schimperi includes specimens with both glabrous and sparsely pilose pedicelled spikelets; these intergrade so gradually that they can scarcely be separated. Are we then justified in maintaining *H. rudis*, with hoary villous pedicelled spikelets, as a separate species? The two species are certainly very closely related, but the ranges of spatheole length, awn length and number of awns per raceme-pair are not coincident, and some support can thus be found for maintaining both species; nor do there seem to be many intermediates between sparsely pilose and hoary villous spikelets. A less worthy consideration is that if *H. schimperi* and *H. rudis* are united, it is difficult to see how a sequence of further unions can logically be avoided.

H. schimperi often has a rather longer scarious rim on its raceme-bases than is found in allied species, but the character is so variable that it is of limited value as an aid to identification.

29. **Hyparrhenia cyanescens** *(Stapf)* Stapf in Prain, Fl. Trop. Afr. 9: 351 (1918).

Cymbopogon cyanescens Stapf in Journ. de Bot. sér. 2, 2: 209 (1909). Type: Mali, Chevalier 2359 (P, holotype; K, isotype).
Andropogon cyanescens (Stapf) A. Chev., Sudania 1: 35 (1911).

Robust perennial, erect or ascending from a short rhizome; culms up to 3 m. high and 8 mm. in diameter, sometimes supported by stilt roots. *Leaf-sheaths* glabrous, sometimes produced at the mouth into oblong auricles up to 6 mm. long; ligules up to 3 mm. long; blades up to 50 cm. long and 3–8 mm. wide, long attenuate at the base, firm, glabrous, greyish-glaucous. *Spathate panicle* diffuse, 30–60 cm. long; spatheoles linear–lanceolate, 3·5–5 cm. long, commonly rolled around the peduncle at maturity, glabrous, glaucous grey or at length becoming tinged with purple; peduncles 1–5 cm. long, usually from half to nearly as long as the spatheole but sometimes shorter than this, pilose with white hairs towards the tip. *Racemes* 1·7–2·5 cm. long, 6–11-awned per pair, lax with internodes 2·5–4 mm. long, exserted from near the end of the spatheole; raceme-bases subequal, about 1 mm. long, sparingly beset with stiff bristles, the tip with a scarious rim up to 0·5 mm. long. *Homogamous spikelets* 6–8 mm. long, glabrous except for the ciliolate margins. *Sessile spikelets* 4·5–6 mm. long, glabrous to sparsely pubescent, steely grey at maturity; callus oblong to cuneate, 0·5–1·2 mm. long, subobtuse; awn 2·8–3·4 cm. long. *Pedicelled spikelets* 6–8 mm. long, glabrous to hispidulous, terminating in a bristle 1–2 mm. long.

MINOR VARIATIONS. The raceme-bases are seldom conspicuously hairy, and may occasionally be glabrous. Another oddity is an awnless specimen (*Audru* 738) from Chad.

CHROMOSOME NUMBER. Nigeria, 2n = c. 36 *Shika* B96/63/OP!.

Root tip preparations were not sufficiently clear to permit an accurate chromosome count, but at least the rough count serves to indicate ploidy level. Unfortunately the plant died before better preparations could be attempted.

HABITAT. Moist alluvial ('fadamma') soils.

DISTRIBUTION. West Africa; with isolated occurrences on the western coast south of the equator (Map 8, p. 109).

GAMBIA. Without locality, *Ruxton* 32.

MALI. Myomina to Koulikoro (Oct.), *Chevalier* 2359.

GUINEA. Without locality, *Scaetta* 3152; Kouroussa (Sept.), *Pobéguin* 508; Kankan (Oct.), *Duong*.

SIERRA LEONE. Musaia (Dec.), *Deighton* 4440.

IVORY COAST. Boundiali (Oct.), *Boudet* 3312.

GHANA. Sawla to Wa (Oct.), *Rose Innes* GC 32394; Bole (Oct.), *Rose Innes* GC 32407; Damongo (Mar.), *J. K. Morton* GC 8670; Gushiege to Yendi (Nov.), *Rose Innes* GC 32501; Yendi to Zabzugu (Nov.), *Rose Innes* GC 32471; Tamale to Salaga (Oct.), *Rose Innes* GC 32441.

NIGERIA. Maska (Nov.), *Thatcher* S506; Zaria (Oct.), *Freeman* S154; Anara F.R. (Oct.), *Keay* in FHI 5490; Biu (Oct.), *de Leeuw* 1328a; Bida (Dec.), *Hubbard* S586; Ilorin (Oct.), *J. F. Ward* 39; Olokemeji (Nov.), *Clayton* 568.

REPUBLIC OF CONGO. Madingou (May), *Koechlin* 2677, 2678.

CONGO REPUBLIC. Leopoldville, Kimpese (June), *Compère* 2113.

ANGOLA. Cassualala (July), *Gossweiler* 8392.

It is clear from the scatter diagram (Fig. 25, p. 104) that the West African specimens do not belong to the same population as *H. schimperi*, indeed the difference between them is probably greater than that between members of the *H. schimperi, H. variabilis, H. cymbaria, H. formosa* series. Yet, as so often happens in this genus, it is remarkably difficult to find unambiguous characters for separating the species, owing to the confounding effect of the wide range of variation within a single panicle. Perhaps the most reliable discrimination is provided by the intuitive synthesis performed by the eye, which separates the racemes of *H. schimperi* enclosed within broad russet or purplish spatheoles, from those of *H. cyanescens* exserted from narrow spatheoles of sombre grey.

H. cyanescens is more easily distinguished from *H. dregeana* by its longer awns, and by the noticeably longer internodes of the lax racemes.

30. **Hyparrhenia papillipes** (*Hochst. ex A. Rich.*) *Anderss. ex Stapf* in Prain, Fl. Trop. Afr. 9: 347 (1918).

Andropogon papillipes Hochst. ex A. Rich., Tent. Fl. Abyss. 2: 460 (1851); Hack. in DC., Monogr. Phan. 6: 620 (1889). Type: Ethiopia, *Schimper* 1055 (K, isotype).

Sorghum papillipes (Hochst. ex A. Rich.) Kuntze, Rev. Gen. Pl. 2: 792 (1891).

Cymbopogon papillipes (Hochst. ex A. Rich.) Chiov. in Monogr. Rapp. Colon., Roma 24: 66 (1912), *in obs.*

Hyparrhenia lintonii Stapf in Prain, Fl. Trop. Afr. 9: 350 (1918). Type: Kenya, *Linton* 124 (K, holotype).

Slender tufted perennial arising from a short underground rhizome; culms much branched below, thin and wiry, 30–100 cm. high and 1–2 mm. in diameter at the base, erect or ascending, usually clad in cataphylls at the base.

Leaf-sheaths glabrous or the basal sheaths pubescent to tomentose; ligule 0·5 mm. long; blades short, 5–15 cm. long and 2–4 mm. wide, glabrous, light green to glaucous, tapering to a setaceous point. *Spathate panicle* rather scanty, of 2–10 raceme-pairs; spatheoles linear-lanceolate, 4–7 cm. long, rolled around the peduncle at maturity, glabrous; peduncles flexuous, 3–8 cm. long, from three quarters to rather longer than the spatheole, pilose with yellowish hairs near the tip. *Racemes* tardily deflexed, 2–4 cm. long, 9–19-awned per pair, white hairy, long exserted; raceme-bases subequal, the upper 1–2 mm. long and becoming subterete above, the rim produced into a scarious appendage; appendage narrowly oblong to linear, (0.2–)0·5–1·2(–1·7) mm. long, purple. *Homogamous spikelets* 5·5–7 mm. long, villous. *Sessile spikelets* 5–6 mm. long, purplish, silky villous with white hairs; callus cuneate, subacute, 0·7–1 mm. long; awn 2–3 cm. long. *Pedicelled spikelets* 6–9 mm. long, villous, awnless or rarely with a short awn-point up to 1·5 mm. long; pedicel tooth subulate, (0·2–)0·4–0·8(–1·5) mm. long.

MINOR VARIATIONS. The spikelets are usually densely silky hairy, but may be merely sparsely villous. In one specimen that I have seen (*Gillet* 14239 from Mega, Ethiopia) they are glabrous. The upper raceme-base can be up to 2·5 mm. long, and in a few specimens (notably *Bogdan* 427 & 3479 from Kenya) it is almost glabrous.

HABITAT. A species of rocky hill slopes at 1600–2600 m., forming cushions or scrambling among other grasses.

DISTRIBUTION. Confined to the highlands of Ethiopia and Kenya, apart from limited records beyond the frontiers of the latter (Map 10). There is also a single unlocalized specimen from Madagascar.

CONGO REPUBLIC. Nioka (Aug.), *Taton* 1489.

ETHIOPIA. Adowa (Nov.), *Schimper* 1055; Begemder, *Schimper* 1033; Gondar, Fenter (Oct.), *Chiovenda* 2311, 2318; Gondar, Azozo (Sept.), *Chiovenda* 2007.

UGANDA. Rom Mt. (June), *Liebenberg*; Moroto Mt. (Sept.), *J. G. Wilson* 201; Napak (June), *Eggeling* 5891.

KENYA. Kulal (Oct.), *Bally* 5577; Marsabit (Feb.), *Gillett* 15094; Nakuru to Eldame Ravine (Sept.), *Bogdan* 2069; Meru (Dec.), *Bogdan* 2723; Narok (July), *Glover & Samuel* 3027; Nairobi (Feb.), *Linton* 124; Machakos, Kitandi (Jan.), *Clayton* 40.

TANZANIA. Serengeti central plain, NE. side (June), *Greenway & Turner* 10690; Arusha (Feb.), *Haarer* A56; Kilimanjaro (Jan., July), *Haarer* 1695, *Greenway* 6701; Mbulumbulu (July), *Greenway* 6767.

MADAGASCAR. Without locality, *Warpur* 559.

H. papillipes is perhaps the most distinct of the species in sect. *Pogonopodia*; it may usually be recognized at sight by its remarkably slender stature with short leaf-blades, by its long narrow spatheoles, frequently with terminally exserted peduncles, and by the relatively large number of sessile spikelets in the raceme. Other distinctive features are the little oblong appendage on the raceme-bases and the subulate pedicel tooth, which help to separate it from *H. dregeana* and *H. collina* with which it intergrades.

The presence of a fairly long appendage might seem to place it in sect. *Hyparrhenia*, but it has nothing else in common with this section whose species usually have glabrous spikelets and, with the exception of *H. coleotricha*, are 2–4-awned per raceme-pair. *H. coleotricha* is a robust annual with longer awns, fewer spikelets, and aristate pedicelled spikelets, so there is little chance of confusion. On the other hand the racemes of *H. papillipes* are typical of other hairy

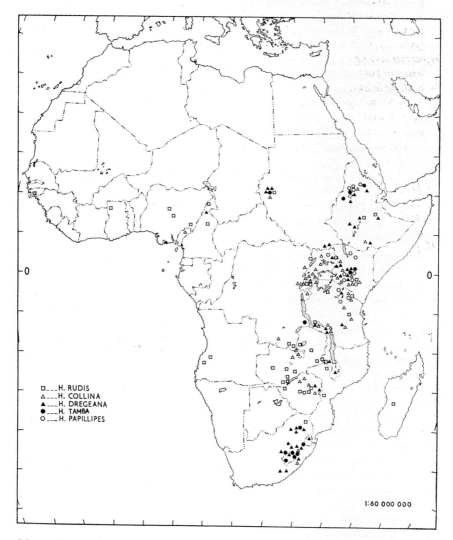

MAP 10. Distribution of *Hyparrhenia rudis, H. collina, H. dregeana, H. tamba* and *H. papillipes*.

species in sect. *Pogonopodia*, from which, indeed, they cannot always be separated with certainty.

The long sparsely hairy raceme-bases displayed by a few specimens are more typical of sect. *Polydistachyophorum*. However such specimens possess the characteristic vegetative habit of *H. papillipes*, and their curious raceme-base seems to be no more than the extreme expression of a character whose normal range of variation is entirely typical of sect. *Pogonopodia*.

31. **Hyparrhenia dregeana** (*Nees*) *Stapf ex Stent* in Bothalia 1: 249 (1923).

Andropogon dregeanus Nees, Fl. Afr. Austr. 1: 112 (1841); Stapf in Thistleton-Dyer, Fl. Cap. 7: 359 (1898). Type: South Africa, *Drège*.

A. pilosissimus Hack. in DC., Monogr. Phan. 6: 690 (1889). Type: South Africa, *Rehmann* 7109 (Z, holotype).

A. auctus Stapf in Thistleton-Dyer, Fl. Cap. 7: 357 (1898). Type: South Africa, *Zeyher* 1799 (K, lectotype).

Hyparrhenia elongata Stapf in Prain, Fl. Trop. Afr. 9: 343 (1918). Type: Ethiopia, *Schimper* 1006 (K, lectotype).

H. phyllopoda Stapf, *l. c.*: 346 (1918). Type: Ethiopia, *Drake-Brockman* 151, 152 (both K, syntypes).

H. aucta (Stapf) Stapf ex Stent in Bothalia 1: 249 (1924).

Cymbopogon micratherus Pilger in Not. Bot. Gart. Berlin 10: 596 (1929). Type: Kenya, *Fries* 1459 (UPS, holotype).

Hyparrhenia micrathera (Pilger) Pilger ex Peter in Fedde, Rep. Sp. Nov., Beih. 40 (1): 379 (1936).

H. brachychaete Peter in Fedde. Rep., Beih, 40 (1): 376, & Anh. 121, t. 78 (1936). Type: Tanzania, *Peter* 44224 (B, destroyed during second World War).

H. subaristata Peter, *l. c.*: 375 & Anh. 120, t. 77 (1936). Type: Tanzania, *Peter* 43327 (B, destroyed during second world war).

H. pilosissima (Hack.) J. G. Anderson in Bothalia 9: 130 (1966).

Robust densely caespitose perennial; culms simple or scantily branched, 1·5–2 m. high or more, up to 4 mm. in diameter at the base. *Leaf-sheaths* glabrous expect for the silky pubescent basal sheaths; ligule rounded, up to about 4 mm. long; blades mostly up to 60 cm. long and 3–8 mm. wide (rarely narrow and rolled), stiff, glaucous, glabrous, scaberulous on the nerves and harsh to the touch, but sometimes soft and green. *Spathate panicle* narrow, fairly dense, 20–50 cm. long; spatheoles narrowly lanceolate, 2·5–5 cm. long, glabrous or weakly hirsute (rarely the whole panicle copiously villous), becoming reddish brown; peduncles 1·5–5 cm. long, from half to rather longer than the spatheole, pilose towards the tip with yellowish hairs. *Racemes* tardily deflexed, 2–3 cm. long, 10–25-awned per pair, densely spiculate with internodes 1·5–2 mm. long, glabrescent to villous, usually exserted terminally; raceme-bases subequal, the upper 1–1·5 mm. long, the tip with a scarious rim and usually produced into a definite oblong appendage up to 0·5 mm. long. *Homogamous spikelets* 5–7 mm. long, villous or rarely glabrous. *Sessile spikelets* 4–5 mm. long, densely villous to hispidulous (rarely glabrous), yellowish green to light brown or purple; callus cuneate, subacute to narrowly obtuse, 1 mm. long; awn (1–)8–20(–28) mm. long. *Pedicelled spikelets* 5–6 mm. long, similar to the sessile in indumentum and colouring, muticous or with a short awn-point up to 1·5 mm. long.

MINOR VARIATIONS. The awn is usually over 8 mm. long, but plants may be found with some of the spikelets almost awnless. They are always mixed with shortly awned spikelets, and seem to differ in no essential respect from similar panicles with normal awns.

Another oddity shown by a few plants is a tendency for the spatheoles in parts of the inflorescence to become much reduced and sessile within the spathe. The apparent spatheole (really a spathe) thus subtends several racemes-pairs cach exserted on a long peduncle.

The spikelets may be infected by the smut *Sphacelotheca barcinonensis* Riofrio.

CHROMOSOME NUMBER. South Africa, 2n = 20 (Krupko, 1953); 2n = 20 (Krupko, 1955); 2n = 20 (de Wet, 1954); 2n = 20 (de Wet, 1958). Kenya, 2n = 20 *Bogdan* 5158!.

HABITAT. In Kenya and Ethiopia it occurs at 1700–3000 m. in open savanna and stony hillsides. In southern Africa it is found at lower altitudes (1000–2000 m.) in open grasslands, favouring streamside sites or the drier soils around vleis.

DISTRIBUTION. The eastern half of Africa from Cape Province to Ethiopia, but most frequent at the northern and southern extremities of this range (Map 10, p. 123).

CAMEROUN. Garoua (Oct.), *Koechlin* 7342.

SUDAN. Jebel Marra (Dec.), *Lynes* 179; Jebel Marra, Wadi Amer (Dec.), *J. K. Jackson* 3385; Jebel Marra, Suni (Dec.), *J. K. Jackson* 3295; Jebel Marra, Gollol (June), *Blair* 36; Imatong Mts., Kipia to Itobol (Sept.), *Myers* 13498; Didinga Mts., Nagichot (Sept.), *Myers* 13472.

ETHIOPIA. Tigré, *Schimper* 1006; Gondar (Oct.), *Chiovenda* 2428; Addis Ababa (Nov.), *Mooney* 5062; Upper Godeb gorge (Aug.), *Hiller & Lythgoe* 504; Jimma (Oct.), *R. B. Stewart* 184; Soddu (Oct.), *Mooney* 6104, *Scott* 70; Mukafilo, *Drake-Brockman* 151, 152; Gara Mullata (Nov.), *Burger* 2350.

UGANDA. Moroto Mt. (Jan.), *J. G. Wilson* 641.

KENYA. Mount Elgon (July), *Pole-Evans & Erens* 1509; Kitale (Jan.), *Bogdan* 4112; Thompson's Falls (Sept.), *Barney* 1156; Nanyuki (June, Sept., Dec.), *Moreau* 68, *Bogdan* 2756, *Geesteranus* 6096; Elburgon to Turi (June), *Bogdan* 1663; Naro Moru, *Muthambi* 5158; Nakuru (Sept.), *Hitchcock* 25116; Gilgil (Jan.), *Bogdan* 4764, *Dowson* 609; Kinangop (Sept.), *Turner* 1574; Ngong Hills (June), *Bogdan* 3058.

TANZANIA. Ngorongoro (June), *Pole-Evans & Erens* 967; Sumbawanga, Luiche valley (Apr.), *Michelmore* 1034; Mbeya (Mar.), *St. Clair Thompson* 801; Tukuyu, Kiwira F.R. (June), *Mgaza* 476; Dabaga (Oct.), *Geilinger* 1979; Mufindi (Oct.), *R. M. Davies* B27.

MALAWI. Lilongwe to Kasungu (Aug.), *Wiehe* N180; Zomba (Aug.), *Wiehe* N614, *G. Jackson* 140.

RHODESIA. Macheke (May), *Fitt* 156; Stapleford F.R. (June), *Gilliland* 275; Nuza (June), *Gilliland* 541.

SWAZILAND. Mabane (May), *Pole-Evans* 3433.

LESOTHO. Leribe (Feb.), *Phillips* 4455, *Dieterlen* 206; Mpitis to Tsoelike (Mar.), *Ruddock*.

SOUTH AFRICA. Transvaal: Kaalfontein (Apr.), *Louw* 982; Krugersdorp (Feb.), *Rodin* 3822; Pienaar's River (Mar.), *Pole-Evans* 703; Pretoria (Feb.), *Burtt-Davy* 1694; Belfast (Mar.), *Pole-Evans* 3687. Orange Free State: Kroonstad (Apr.), *Chennells* 31; Bethlehem (Jan.), *Liebenberg* 6820; Memel (Apr.), *Acocks* 12564; Thaba Nchu (Feb.), *Zeyher* 1799. Natal: Charlestown (Mar.), *J. M. Wood*; Eastcourt (Apr.), *Acocks* 11477, *Rehmann* 7310; Nottingham Road (Mar.), *McClean* 30415; Boston (Feb.), *D. Edwards* 3236; Pietermaritzburg (Mar.), *Moll* 739. Cape Province: Queenstown (Apr.), *Galpin* 2378; Kei River to Gekau, *Drège*; Fort Beaufort (Apr.), *Story* 2230.

H. dregeana may usually be recognized from its racemes, with their numerous, rather densely packed, spikelets and short awns; moreover the racemes tend to be further exserted from the spatheole than is usual in this section. The culms

9

are stout, densely caespitose, and clad below in pubescent to tomentose basal sheaths. The latter appear to provide a reliable diagnostic character, but it must be admitted that the perennating parts of the other, non-tussocky, species are poorly collected and may not constitute an adequate sample. Other useful, but inconstant, characters are the muticous pedicelled spikelets and the small scarious appendage of the raceme-base. The latter is usually more distinct than in the other hairy species (except *H. papillipes*), but the character is so much the prey of subjective interpretation that I have reluctantly abandoned it for diagnostic purposes.

The species is more variable than any other in the section, particularly in such characters as spikelet indumentum and colour, spatheole indumentum, and the colour, stiffness and breadth of the leaves. Infra-specific taxa could be based upon all these features, but since they are not clear-cut, uncorrelated with other characters and, at present, unsupported by cytogenetic or ecological evidence, they would appear to be quite unhelpful. It is true that certain trends can be discerned—specimens from the north-east tend to have shorter spatheoles, fewer spikelets and longer awns than those from the south—but it is apparently impossible to disentangle them in a meaningful manner.

The racemes are typically villous, but may be rather sparsely so, and are occasionally glabrous. However the measured characters of the glabrous specimens display the same distribution pattern as those of the hairy form, and, while approaching some of the wholly glabrous species, they differ in one or more essential characters and so cannot be happily accommodated with them.

32. **Hyparrhenia tamba** (*Hochst. ex Steud.*) *Anderss. ex Stapf* in Prain, Fl. Trop. Afr. 9: 336 (1918).

Andropogon tamba Hochst. ex Steud., Syn. Pl. Glum. 1: 385 (1854). Type: Ethiopia, *Schimper* 911, 937 (both K, isosyntypes).

A. lepidus var. *tamba* (Hochst. ex Steud.) Hack. in DC., Monogr. Phan. 6: 625 (1889).

Cymbopogon tamba (Hochst. ex Steud.) Rendle in Journ. Linn. Soc., Bot. 40: 227 (1911).

Hyparrhenia glauca Stent in Bothalia 1: 251 (1923). Type: South Africa, *Stent* H21425 (K, isotype).

Stout perennial, forming a tussock without conspicuous stilt roots; culms 1–2 m. high, up to 4 mm. in diameter at the base. *Leaf-sheaths* silky pubescent at the base; ligule rounded, up to 4 mm. long; blades up to 80 cm. long and mostly 3–7 mm. wide, glabrous, stiff, harsh, scabrid on the margins and scaberulous on the nerves, strongly glaucous, rarely soft and green. *Spathate panicle* linear-oblong, loose, 30–60 cm. long; spatheoles narrowly lanceolate, 2·6–4 cm. long, glabrous or with a few scattered hairs, turning glaucous brown; peduncles 2–3 cm. long, $\frac{1}{2}$–$\frac{3}{4}$ as long as the spatheole, pilose towards the tip with white hairs. *Racemes* 1·5 cm. long, 5–8-awned per pair, villous, internodes 2–3 mm. long, exserted laterally on the curved peduncle; raceme-bases subequal, up to 1 mm. long, the tip produced into a lobed scarious rim 0·2–0·5 mm. long. *Homogamous spikelets* 8–9 mm. long, villous. *Sessile spikelets* 5 mm. long, villous with white hairs, becoming dark purplish grey at maturity; callus cuneate, subacute, o·5 mm. long; awn 16–25 mm. long. *Pedicelled spikelets* 7–8 mm. long, villous, awnless or with a bristle up to 2 mm. long.

ILLUSTRATION. Stent in Bothalia 1: 250, t. 3 (1924).

HABITAT. Roadsides and along streams.

DISTRIBUTION. Mainly in South Africa, but similar specimens have been collected in NE. tropical Africa and the species is probably distributed throughout the area of *H. dregeana* (Map 10, p. 123).

CONGO REPUBLIC. Marungu, Ndawa (June), *Dubois* 1209.
SUDAN. Jebel Marra (Feb., Mar.), *Blair* 230, 242, *J. K. Jackson* 3870.
ETHIOPIA. Adowa, Mt. Scholoda (July), *Schimper* 937; Gennia (Dec.), *Schimper* 911; Bellaka (Nov.), *Schimper* 469; Gondar (Sept., Oct.), *Chiovenda* 1828, 2368, 2609, 2626; Gojjam, Kobalit (Aug.), *Lythgoe* 719.
KENYA. Mt. Kenya, *Hutchins* 377; Nyeri (Dec.), *Fries* 146.
LESOTHO. Likhoele (Apr.), *Dieterlen* 1104.
SOUTH AFRICA. Transvaal: Pretoria (Feb., Mar.), *Stent* H21425, *Schweic-kerdt* 1772, *C. A. Smith* 6094; Piet Retief (May), *Pole-Evans* 3778. Orange Free State: Ladybrand (May), *Bakker* 187. Natal: Mont-aux-Sources (Apr.), *Mogg* H 20632; Cathedral Peak (Apr.), *Killick* 2359; Eastcourt (Feb., May), *West* 249, *Acocks* 11492, 11499.

The distribution of the measured characters of *H. tamba* overlaps, but does not coincide with, that of *H. collina* and *H. dregeana*. Indeed, the racemes might be mistaken for *H. collina* at first sight, but *H. tamba* is distinguished by its more robust habit, pubescent basal sheaths, and appreciable appendage on the raceme-base, characters which it shares with *H. dregeana*. From the latter it is distinguished by possessing fewer awns per raceme-pair, and might be regarded as a mere extension of its range were it not for the absence of the characteristically short rhachis internodes and rather long peduncles found in *H. dregeana*. For this reason I have decided to maintain both species, despite their suggestively similar geographical distribution.

Soft, green and flexuous leaved specimens from Ethiopia fall into the region of overlap between *H. dregeana*, *H. tamba*, *H. collina* and *H. rudis*. These species are indistinctly separated in this area, and the meagre number of collections available provides an inadequate sample of the merging populations, so that it is difficult to know how they should be treated.

33. **Hyparrhenia umbrosa** (*Hochst.*) *Anderss. ex W. D. Clayton*, comb. nov.; Anderss. in Schweinf., Beitr. Fl. Aethiop.: 316 (1867), *nom. invalid.*

Andropogon umbrosus Hochst. in sched., Schimp. Iter Abyss. 2: 1116 (1842); A. Rich., Tent. Fl. Abyss. 2: 467 (1851). Type: Ethiopia, *Schimper* 1116 (P, holotype).
A. lepidus subvar. *umbrosus* (Hochst.) Hack. in DC., Monogr. Phan. 6: 625 (1889).
Cymbopogon umbrosus (Hochst.) Pilger in Not. Bot. Gart. Berlin 10: 598 (1929).

Robust perennial arising from a slender rhizome; culms 1·3–2 m. high, slender and rambling below, then becoming thicker, up to 6 mm. in diameter, supported by stilt roots and erect. *Leaf-sheaths* glabrous; ligule 2 mm. long, truncate; blades up to 60 cm. long and 12 mm. wide, glabrous. *Spathate panicle* narrowly oblong, 20–30 cm. long and 4–8 cm. wide, dense, decompound; spatheoles narrowly ovate, 1·2–2·3 cm. long, glabrous or weakly pilose along the margin, russet brown; peduncles 0·3–1·3 cm. long, $\frac{1}{4}$–$\frac{1}{2}$ as long as the spatheole, pilose above with white hairs. *Racemes* 1 cm. long, 4–6-awned per pair, laterally exserted; raceme-bases subequal, 0·5 mm. long, the tips truncate. *Homogamous spikelets* 5–6 mm. long, villous. *Sessile spikelets* 4 mm. long, villous with white hairs; callus oblong, 0·4 mm. long, rounded at the tip; awn 0·7–1·3 cm. long. *Pedicelled spikelets* 5–6 mm. long, villous with white hairs, muticous or mucronate.

HABITAT. Highland grassland at 1700–2300 m., by roadsides and in old cultivation.

DISTRIBUTION. Widely scattered in Africa, but not common (Map 9, p. 112).

NIGERIA. Mambila plateau, Mandaga (Jan.), *Latilo & Daramola* FHI 28995.

CAMEROUN. Nkambe (Feb.), *Charter* FHI 36983; Santa (Jan.), *Brunt* 908; Bambui (Dec.), *Pedder* 4; Djottin (June), *Brunt* 591; Bamenda (June), *Brunt* 1176; Djang (Mar.), *Unwin* 221.

CONGO REPUBLIC. Dubie (May), *Bredo* 4872.

SUDAN. Imatong, Lomuleug (Dec.), *A. S. Thomas* 1806.

ETHIOPIA. Gennia (Dec.), *Schimper* 584; Mt. Scholoda (Dec.), *Schimper* 1116.

UGANDA. Fort Portal (July), *Thornton* 56.

KENYA. Narok, Olulunga (June), *Glover, Gwynne & Samuel* 1592.

TANZANIA. Kibondo (Aug.), *Bullock* 3083; Mufindi (Oct.), *R. M. Davies* B30.

SOUTH AFRICA. Natal: Newcastle, Mount Prospect (May), *Pole-Evans* 3775.

The most characteristic feature of *H. umbrosa* is its oblong callus with a broadly rounded tip; this is coupled with short cymbiform spatheoles and an initially rambling culm (compare the similar, but initially erect, culms of *H. rudis*). Taken together they enable *H. umbrosa* to be distinguished although, as is usual in this section, the characters tend to intergrade with those of neighbouring species.

The wide geographical distribution and obvious similarity to *H. cymbaria* and *H. formosa* suggest that *H. umbrosa* may be no more than a hairy variant. But in that case its apportionment between the two glabrous species would present an awkward problem and, in the absence of definite evidence, it seems more practical to retain it as a separate species.

34. **Hyparrhenia rudis** *Stapf* in Prain, Fl. Trop. Afr. 9: 344 (1918). Type: Angola, *Gossweiler* 4151 (K, lectotype).

H. acutispathacea var. *pilosa* Bamps in Bull. Jard. Bot. Brux. 25: 392 (1955). Type: Congo Republic, *Quarré* 1727 (K, isotype).

Robust coarsely tufted perennial arising from short slender rhizomes clad in white cataphylls; culms stout, erect, 2–3 m. high and up to 8 mm. in diameter at the base, usually supported by stilt roots. *Leaf-sheaths* glabrous; ligule 2–10 mm. long, acute or truncate; blades 30–60 cm. long and 3–18 mm. wide, pale green to glaucous, stiff, harsh, glabrous, scabrid on the margins and scabrid or scaberulous on the nerves. *Spathate panicle* narrowly oblong, 30–50 cm. long, loose or somewhat contracted, decompound; spatheoles narrowly lanceolate, 2·5–4 cm. long, glabrous or sometimes weakly pilose, turning reddish brown at maturity; peduncles typically 1–2 cm. long, $\frac{1}{3}$–$\frac{1}{2}$ as long as the spatheole but sometimes almost as long, pilose towards the tip with off-white hairs. *Racemes* about 1·5 cm. long, 4–7-awned per pair, laterally exserted; raceme-bases subequal, the upper 1·5 mm. long, the tips produced into a scarious lobe 0·2–0·5 mm. long. *Homogamous spikelets* 5–9 mm. long, villous. *Sessile spikelets* 5–6 mm. long, pale or reddish brown, silky villous with white hairs; callus narrowly oblong, 0·6 mm. long, narrowly obtuse at the tip; awn 2·2–4 cm. long. *Pedicelled spikelets* 6–7 mm. long, villous with white hairs, terminating in a bristle 2–6 mm. long, or sometimes merely acuminate.

MINOR VARIATIONS. Two specimens from Zambia (*Verboom* 1370 & *Mitchell* 25/33) should apparently be included in this species, but the spikelet indumentum is definitely fulvous, suggesting introgression from *H. rufa*.

HABITAT. Savanna grasslands, favouring damp or alluvial soils, and riverine sites or dambo margins.

DISTRIBUTION. Best developed in south tropical Africa, but extending northward through the Congo/Tanzania border to Nigeria and Jebel Marra (Map 10, p. 123); also in Madagascar.

GHANA. Sawla to Wa (Oct.), *Rose Innes* GC 32395.

NIGERIA. Zaria (Jan.), *Bumpus* B6; Vom (Oct.), *Gambles* 41; Tiba Plateau (Dec.), *Peter & Tuley* 52.

CAMEROUN. Homsiki (Nov.), *Vaillant* 2478; Ngaundéré (Oct.), *Koechlin* 7435.

CONGO REPUBLIC. Tumbwe (Apr.), *Symoens* 8640; Kipila (May), *Quarré* 1727.

RWANDA. Parc National Kagera (Apr.), *Troupin* 6836; Mt. Lutare (Jan.), *Lebrun* 9598; Kigali to Kakisumba (Jan.), *Troupin* 5794.

SUDAN. Jebel Marra (Oct., Dec.), *Blair* 180, 181, *J. K. Jackson* 3294, *Pettet* J37, J47.

ETHIOPIA. Gondar (Oct.), *Chiovenda* 2460, 2606; Asoso (Sept.), *Chiovenda* 2029; Majjio, Koka (Oct.), *Mooney* 7535; Dira Dawa (Nov., Dec.), *IECAMA* F77, G39.

UGANDA. Kamalinga (Dec.), *Brown* 20.

KENYA. Thika (June), *Bogdan* 761.

TANZANIA. Mwanza, Ngudu (July), *Lewys Lloyd* 15; Sumbawanga, Luiche (Apr.), *Michelmore* 1035; Sumbawanga (Apr.), *Vesey-FitzGerald* 1710; Iheme (Oct.), *McGregor* 22; Mpwapwa (Apr.), *Staples* 217 of 1934.

MOZAMBIQUE. Angónia (May), *Mendonça* 4191; Révué valley (Apr.), *Vasse* 209.

MALAWI. Fort Manning, Nyoka (Apr.), *G. Jackson* 715a, 774, 775, 776; Lilongwe (Apr.), *G. Jackson* 473.

ZAMBIA. Samfya (June), *Seagrief* 3110; Chilonga (May), *Lusaka Nat. Hist. Soc.* 265; Fort Jameson (Apr.), *Verboom* 702; Mumbwa (Mar.), *van Rensburg* 1722; Pemba (May), *Trapnell* 2027; Choma (Apr.), *van Rensburg* 2089; Kalomo to Dundumwenzi (Feb.), *Mitchell* 24/87; Siantambo (Mar.), *Mitchell* 5/96.

RHODESIA. Victoria Falls (Aug.), *C. Sandwith* 15; Salisbury (Apr.), *Kimpton*; Salisbury, Gwebi farm (Mar.), *Barnes* 1; Hartley, Poole farm (Apr.), *Hornby* GH 14905; Hartley, Mandoro Reserve (June), *R. M. Davies* GH 32698; Hartley, Umsweswe River (May), *Rattray* 1464.

ANGOLA. N'jaia (Apr.), *Gossweiler* 3121; Princeza Amelia (May), *Gossweiler* 3919; Munongue (Apr.), *Gossweiler* 3129, 4151.

MADAGASCAR. Mandoto (Dec.), *Decary* 15252; Sakaleona valley (June), *Decary* 14344.

SOUTH AFRICA. Transvaal: Politsi (Apr.), *Scheepers* 242.

H. rudis perennates by means of a short rhizome, from which arise slender shoots in a loose clump. Unlike *H. collina*, whose culms are uniformly slender, the first few internodes of the stem become successively thicker, the augmented culm relying upon supplementary stilt roots for nourishment and support. Compare this also with the growth-form of *H. dregeana* in which the culms, invested below by the ragged remnants of old sheaths, are compacted into a tussock. Apart from the difference in growth habit, *H. rudis* tends to have a longer awn, and a longer bristle on its pedicelled spikelet than *H. collina* and *H. tamba*, with which it is obviously very closely related.

35. **Hyparrhenia collina** (*Pilger*) *Stapf* in Prain, Fl. Trop. Afr. 9: 337 (1918).

Andropogon collinus Pilger in Mildbr., Wiss. Ergebn. Deutsch. Zentr.-Afr. Exped. 2: 43 (1910), *non* Lojac. (1909). Types: Rwanda, *Exped.* No. 375, and Tanzania, *Volkens* 352.

Cymbopogon collinus Pilger in Engler, Bot. Jahrb. 54: 287 (1917). Based on *Andropogon collinus* Pilger, *non* Lojac.

C. scabrimarginatus De Wild. in Bull. Jard. Bot. Brux. 6: 20 (1919). Type: Congo Republic, *Ringoet* 11 (K, isotype).

Andropogon scabrimarginatus De Wild., *l. c.* (1919), *in synon.*

Hyparrhenia scabrimarginata (De Wild.) Robyns, Fl. Agrost. Congo Belge 1: 184 (1929).

Loosely clumped perennial arising from a short rhizome clad in white cataphylls; culms erect or ascending, wiry and slender, 30–130 cm. high and 1–3 mm. in diameter at the base, without stilt roots. *Leaf-sheaths* glabrous; ligules 1–2 mm. long; blades up to about 30 cm. long and 2–5 mm. wide, firm but scarcely rigid, scabrid on the margins and scaberulous on the nerves. *Spathate panicle* narrow, 15–40 cm. long, scanty; spatheoles narrowly lanceolate, 2–4 cm. long, glabrous or weakly hirsute, turning reddish brown; peduncles 1–2·5 cm. long, rather more than half as long as the spatheole, pilose with yellowish hairs above. *Racemes* 1–2 cm. long, 4–7-awned per pair, hoary or brownish grey, laterally exserted; raceme-bases subequal, the upper 1–1·5 mm. long, with an inconspicuous scarious rim. *Homogamous spikelets* 5–6 mm. long, villous. *Sessile spikelets* 4·5–5 mm. long, usually dark purple beneath a covering of white hairs; callus cuneate, 0·5 mm. long, narrowly obtuse at the tip; awn 1·5–2·5 cm. long. *Pedicelled spikelets* 4–7 mm. long, villous, terminating in an awn-point 1–3 mm. long.

HABITAT. Alluvial soils, heavy clays, mbugas and seasonally flooded land; but also recorded on upland savannas and hillside grassland.

DISTRIBUTION. Centred on east Africa, but extending north to the Sudan and south to Natal (Map 10, p. 123).

CAMEROUN. Nkambe (June), *Brunt* 977.

CONGO REPUBLIC. Lulenga to Sake (Feb.), *Lebrun* 5032; Beni to Kasindi (Dec.), *Lebrun* 4644; Kivu, Muhambi-Rubengesa, *Scuellu* 1705 bis; Kalulu (May), *Bredo* 2866; Tshinsenda (May), *Ringoet* 11.

RWANDA. Mutara (Mar.), *Troupin* 6741; Nyakayaga (Jan.), *Lebrun* 9462; Parc Nat. Kagera (Jan., Feb.), *Troupin* 5889, 6119; Astrida (Aug.), *Gilon* 145.

BURUNDI. Kitaramuka (June), *Van der Ben* 2109.

SUDAN. Wadi Burora (Jan.), *J. K. Jackson* 2559.

ETHIOPIA. Simien, Agrima, *Schimper* 138; Gondar (Oct.), *Chiovenda* 2370; Asoso, *Chiovenda* 2536; Sidamo, Mogada forest (Jan.), *Mooney* 5480.

UGANDA. Gulu (Nov.), *A. S. Thomas* 4405; Nabuswera (Sept.), *Langdale-Brown* 1530; Kyegegwa (June), *A. S. Thomas* 4136; Bugangari (May), *Purseglove* 3394; Kasheshe (Feb.), *Snowden* 1307; Kambuga (Apr.), *A. S. Thomas* 3796; Lutobo (Dec.), *A. S. Thomas* 1218; Lwasamaire (May, June, Nov.), *Snowden* 1351, *A. S. Thomas* 4188, *H. B. Johnston* 1336; Mbarara (Dec.), *Brockington* 27.

KENYA. Eldoret (Apr., Aug., Sept.), *Williams* 137, *Bogdan* 1914, *Hitchcock* 25018; Narok, Olongeri (Apr.), *Glover, Gwynne & Samuel* 875; Mt. Aberdare (Feb., Dec.), *Fries* 455, 1991; Thika (Jan.), *Bogdan* 1489; Ngong (July), *van Someren* AH 9585; Nairobi, *Dowson* 238, *Hitchcock* 24722, *Bogdan* 308, 407, *McCallum Webster* K180; Langata forest (June), *Bogdan* 754.

TANZANIA. Bukoba, Minziro forest (Jan.), *Procter* 785; Kilimanjaro (Jan.), *Haarer* 1687; Marangu (Jan., Sept.), *Schlieben* 4634, *Hitchcock* 24596, 24669; Kondoa, Sambala (Mar.), *B. D. Burtt* 2037; Iringa (July), *Milne-Redhead & Taylor* 11169; Pangire to Ngongwa R., *Wolfe* 15, 27; Sumbawanga (Apr.), *Phipps & Vesey-FitzGerald* 3323; Mbeya (May), *Milne-Redhead & Taylor* 10053; Rungwe (Mar.), *R. M. Davies* 515; Bundali (May), *Stolz* 1296.

MALAWI. Mperere Mission (Nov.), *G. Jackson* 253.

ZAMBIA. Kafue R. headwaters (June), *Angus* 2356; Munshiwemba (Nov.), *Stohr* 653.

RHODESIA. Miami (Mar.), *Wild* 1694; Urungwe, Msukwe R. (Nov.), *Wild* 4207; Salisbury (Apr., May), *Brain* 4239, *Gilliland* 60, 64; Inyazura (Mar.), *Eyles* 2388, 2396.

SOUTH AFRICA. Natal: Biggarsbergen, *Rehmann* 7116; Umpumulo (Nov.), *Buchanan*.

The distinguishing features of *H. collina* are summarized in the key and the scatter diagram (Fig. 35, p. 169), from which it is apparent that it is rather indistinctly separated from *H. rudis*, *H. dregeana* and *H. tamba*. However, the morphological parameters of the four taxa apparently delineate separate populations to which has been applied, of necessity arbitrarily, the rank of species. From the results of this classification further observations can be derived, which shed some light on the nature of these populations.

The first point is the ambivalent nature of the habitat notes for all four species—mainly moist soils, but sometimes dry savanna. An attempt to recast the analysis failed to find any satisfactory correlation between habitat and morphology, so it can only be assumed that the four species possess some degree of adaptability to their environment. Alternatively this may merely be a reflection of the generally inadequate and confused nature of the ecological notes entered on herbarium labels.

The second observation relates to geographical distribution. All four species are widespread, but *H. dregeana* and *H. tamba* are concentrated in southern and north-eastern Africa, *H. rudis* in Zambia and Rhodesia, and *H. collina* in the East African territories. It seems, therefore, that the morphological populations have a strong geographical bias, which provides additional support for the decision to separate them taxonomically.

IV. Sect. **Hyparrhenia**

Hyparrhenia series *Bracteatae* Stapf in Prain, Fl. Trop. Afr. 9: 293 (1918). Type species: *H. bracteata* (Humb. & Bonpl. ex Willd.) Stapf.
Hyparrhenia sect. *Bracteola* Pilger in Engl., Nat. Pflanzenfam. 14e: 174 (1940), *sine descr. lat.*

Tall annuals or perennials, the former sometimes dwarfed on poor soils. *Leaf-sheaths* sometimes auricled; ligules scarious, 1–20 mm. long; blades linear, petiolate in a few species. *Spathate panicle* usually copious, lax or sometimes dense; spatheoles narrowly lanceolate, dull green becoming russet; peduncles almost as long as the spatheoles (except *H. madaropoda*), pilose near the tip. *Racemes* deflexed (by means of the peduncle in *H. madaropoda*), few-awned, often purplish and surrounded by yellow hairs; raceme-bases subequal, flattened (or rather unequal and the upper subterete in *H. madaropoda* and perennial species), the lower 0·5–1 mm. long, the upper 1–1·5(–3) mm. long, bearded with stiff glassy tubercule-based bristles, the tips oblique with their rim produced on one side into a pallid (annual) or purple (perennial) appendage; appendage scarious, oblong to linear, denticulate, 0·5–4 mm. long (absent in *H. niariensis* var. *macrarrhena*). *Homogamous pairs* 1 at base of lower raceme; spikelets linear-lanceolate, usually glabrous and muticous. *Sessile spikelets* usually narrowly oblong; callus pungent, or sometimes acute in perennial species; upper lemma lobes 0·1–0·5(–1) mm. long; awns prominent, the column pubescent with rufous or fulvous hairs 0·2–0·6 mm. long. *Pedicelled spikelets* linear-lanceolate, usually awned; callus usually absent, sometimes up to 0·3 mm. long and then broad.

The section is distinguished by the heavily bearded raceme-bases and by the presence of an appendage. The latter is absent in *H. niariensis* var. *macrarrhena*, but the variety is so similar to the species in other respects that it can scarcely be separated. The raceme-bases are typically short and flattened, but may be rather longer and subterete in the perennial species. *H. madaropoda* violates several of the diagnostic criteria, but is clearly allied to *H. niariensis* (see discussion of the species).

The section is closely related to sect. *Pogonopodia*, in which a few species (notably *H. papillipes*) also possess a small appendage. In such cases sect. *Hyparrhenia* can readily be distinguished by the few-awned raceme-pairs, and by the lack of boat-shaped spatheoles enclosing the racemes.

This section bears some resemblance to *Hyperthelia*, but that genus is readily separated by the pronounced median groove in the lower glume of the sessile spikelet.

The section may be divided into two parts containing annual and perennial species respectively. Associated with this division is a difference in the texture of the lower glume of the sessile spikelets. In annuals it tends to be pallid, flat-backed, papery and distinctly nerved throughout; whereas in perennials it is usually purplish, rounded, subcoriaceous and indistinctly nerved, except for the middle nerve which is raised between two slight grooves. These differences, though not sharp, suggest that the distinction is more fundamental than the mere difference in habit would suggest.

TABLE 8. Characters of diagnostic value among the annual species of sect. *Hyparrhenia*

	Awns per raceme-pair	Length of sessile spikelet (mm.)	Length of pedicelled spikelet (mm.)	Length of awn on p. spikelet (mm.)	Length of appendage (mm.)	Peduncle hairs (white) (yellow)
H. madaropoda	2	7–9·5	8–9	0–6	0·5–0·7	W
H. coleotricha	4–7	7–9	8–12	6–8	1–2	W
H. confinis var. confinis ⎫ var. pellita ⎭	2	8–10	11–14	10–17	1–2	W
H. confinis var. nudiglumis	2	7–8·5	6–10	9–14	3–4	W
H. niariensis var. niariensis	2(–3)	7·5–11	8–13	2–10	0·5–4	Y(W)
H. niariensis var. macrarrhena	2	8·5–10	12–14	1–9	0–0·2	Y
H. welwitschii	3	5–7	6–8	2–11	0·5–1	WY

Among the annual species *H. madaropoda* is quite distinct. Specimens having 3 awns per raceme-pair (with a few exceptions discussed under *H. niariensis*) form a compact nodum which may be separated as *H. welwitschii*. The remainder of the annual complex has usually been regarded as a single species, *H. confinis*, but it is clear that specimens from the Ethiopian and Congolan regions belong to different taxa. Unfortunately the collections available for study constitute an inadequate sample for a proper understanding of variation in this group, and its treatment must be, to some extent, empirical.

The characters of the species are summarized in table 8. It may be noted that all the species have large ligules and a tendency to auricles.

The perennial species also intergrade, but the difficulties may be relegated to a discussion under each species.

36. **Hyparrhenia madaropoda** *W. D. Clayton*, sp. nov.; affinis *H. niariensi* (Franch.) W. D. Clayton, sed basi racemi superioris longiore et glabriore haud deflexa sine appendice distincta, et pedunculo deflexo unilateraliter barbato differt. Type: Uganda, *A. S. Thomas* 4060 (K, holotype).

Annual; culms 1–3 m. high, stout, erect or the lower internodes geniculately ascending, up to 5 mm. in diameter, supported by stilt roots. *Leaf-sheaths* glabrous, produced at the mouth into auricles up to 10 mm. long; ligule up to 25 mm. long, adnate to the auricles; blades up to 45 cm. long and 13 mm. wide, glabrous, the longer blades narrowed at the base into a short false petiole. *Spathate panicle* copious, about 30 cm. long, becoming rather scanty in the poorer specimens; spatheoles lanceolate, 2·5–4 cm. long; peduncles very short, 3–5 mm. long, densely white bearded on one side near the tip. *Racemes* 1·5 cm. long, 2-awned per pair, laterally exserted by flexion of the peduncle; raceme-bases unequal, the upper 2–2·5 mm. long, subterete, meagrely embellished with stiff bristles, the tip very oblique and produced into a scarious extension 0·5–0·7 mm. long. *Homogamous spikelets* 8–9 mm. long, glabrous. *Sessile spikelets* lanceolate, 7–9·5 mm. long, glabrous or hispidulous; callus 1·5–3 mm. long; awn 4·5–6·5 cm. long. *Pedicelled spikelets* 8–9 mm. long, with or without an awnlet up to 6 mm. long; pedicel tooth obscure.

ILLUSTRATION. Figure 28.

HABITAT. Old farmland and disturbed places.

DISTRIBUTION. On the frontiers between Uganda, Sudan and Kenya (Map 11, p. 136).

SUDAN. Bor to Baidit (Oct.), *M. N. Harrison* 1041; Yei (Sept.), *Myers* 7730, 9457; Torit (Oct.), *J. K. Jackson* 871, 872.

UGANDA. Yumbe (Nov.), *Langdale-Brown* 2375; Madi, Metuli (Nov.), *A. S. Thomas* 4060; Lira, Ngetta Rock, *Fiennes* 19; Karamoja, Kaabong (Aug.), *J. G. Wilson* 1146.

KENYA. Kacheliba (Oct.), *Leippert* 5124.

The 2-awned raceme-pairs and the shape and dimensions of the spikelets of *H. madaropoda* are very similar to those of *H. niariensis*. The conviction that the species should be placed in sect. *Hyparrhenia* is strengthened by the long ligule, the auricles and the tendency to a false petiole, characters which are typical of this section. Indeed there is no species in any other section to which a relationship might reasonably be ascribed.

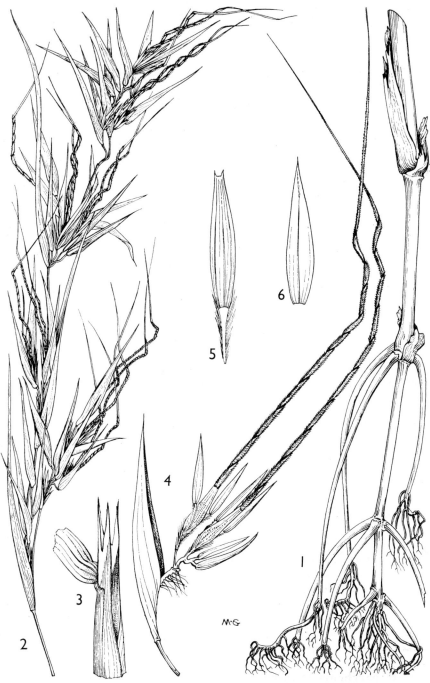

FIG. 28. *Hyparrhenia madaropoda*. **1,** base of plant showing roots and upright habit of stem, ×⅔; **2,** inflorescence, ×⅔; **3,** ligule and auricles, ×2; **4,** raceme-pair, ×2. Sessile spikelet. **5,** lower glume and callus, ×5; **6,** upper glume, ×5. **1** from *A. S. Thomas* 4060, **2–6** from *Wilson* 1146.

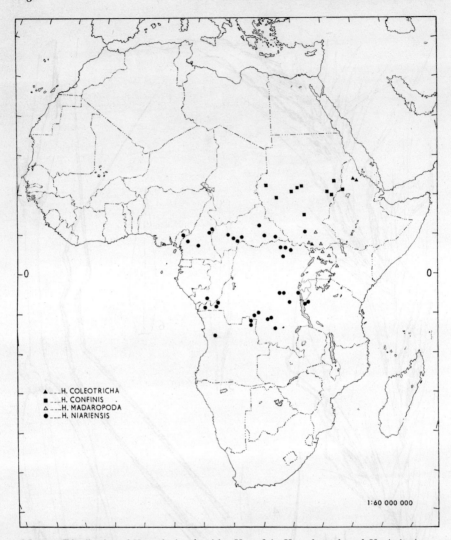

MAP 11. Distribution of *Hyparrhenia coleotricha*, *H. confinis*, *H. madaropoda* and *H. niariensis*

The raceme-bases however are quite atypical in this section, for they are not deflexed, the scarious extension of their very oblique tips can barely be classed as an appendage, and the upper raceme-base is long, subterete and often almost glabrous. However the short peduncle, with its unilateral beard and sharp flexure at maturity, is quite unique in the genus, and the transfer of function from the raceme-bases to this organ may well account for the other peculiarities. I am satisfied that, despite its unusual features, *H. madaropoda* should be classed among the annual species of sect. *Hyparrhenia*.

37. **Hyparrhenia coleotricha** *(Steud.) Anderss. ex W. D. Clayton*, comb. nov.

Andropogon comosus Hochst. *ex* A. Rich., Tent. Fl. Abyss. 2: 461 (1851), *non* Spreng. (1819). Type: Ethiopia, *Schimper* 1458 (K, isotype).

Andropogon coleotrichus Steud., Syn. Pl. Glum. 1: 386 (1854); Hack. in DC.,
   Monogr. Phan. 6: 642 (1889). Based on *A. comosus* Hochst. ex A. Rich.
Sorghum comosum Kuntze, Rev. Gen. Pl. 2: 790 (1891). Based on *Andropogon
   comosus* Hochst. ex A. Rich.
Andropogon anthistirioïdes var. *procerus* Chiov. in Ann. Ist. Bot. Roma 8: 289
   (1908). Types: Eritrea, *Pappi* 1506, 1814, 1869, 6130 (K, isosyntypes).
Hyparrhenia comosa (Kuntze) Anderss. ex Stapf in Prain, Fl. Trop. Afr. 9: 358
   (1918).

Annual; culms 1–2·5 m. high, glabrous or with a ring of fulvous hairs at each
node, the base ascending and supported by stilt roots. *Leaf-sheaths* glabrous or
hirsute; ligule rounded, up to 4 mm. long; blades up to 30 cm. long and 15 mm.
wide, usually glabrous, gradually narrowed at the base. *Spathate panicle* narrow,
loose, 20–30 cm. long; spatheoles 4·5–7 cm. long, glabrous; peduncles half to
nearly as long as the spatheoles, pilose with white hairs near the tip. *Racemes*
2–2·5 cm. long, 4–7-awned per pair, laterally exserted; raceme-bases bearing
white bristles; appendage linear-oblong, denticulate, 1–2 mm. long. *Homogamous spikelets* 10–12 mm. long, glabrous or pubescent below, scaberulous on
the margin, with or without an awnlet up to 3 mm. long. *Sessile spikelets* linear-
lanceolate, 7–9 mm. long, shortly and densely villous with white hairs; callus
1·5–3 mm. long; awn 5–7 cm. long. *Pedicelled spikelets* 8–12 mm. long, more or
less pubescent, terminating in a long fine bristle 6–8 mm. long; pedicel tooth
narrowly triangular, 0·3–0·6 mm. long.

HABITAT. Not known.

DISTRIBUTION. North east tropical Africa, with a single unlocalized specimen
from Tanzania (Map 11).

SUDAN. Yei, Kagelu (Sept.), *Myers* 7732.
ERITREA. Dembelas, Ferfer (Sept.), *Pappi* 6103; Oculé Cusai (Sept.), *Pappi*
1506.
ETHIOPIA. Without locality, *Schimper* 1458.
TANZANIA. Without locality, *Staples* 429 of 1932.

An uncommon and imperfectly known species, distinguished from the
other annual species by the possession of 4–7 awns per raceme-pair. It is not
entirely homogeneous and this, together with the diffuse geographical distribution, suggests that it may be an artificial grouping of several-awned variants
from the other species in the section. However, a glance at the scatter diagram
(Fig. 29, p. 138) shows that it is by no means clear to which of these species the
several-awned specimens should be related, and for the time being it seems
preferable to keep them apart.

38. **Hyparrhenia confinis** (*Hochst. ex A. Rich.*) *Anderss. ex Stapf* in Prain, Fl. Trop. Afr. 9: 353 (1918).

Andropogon confinis Hochst. ex A. Rich., Tent. Fl. Abyss. 2: 461 (1851); Hack.
   in DC., Monogr. Phan. 6: 640 (1889). Type: Ethiopia, *Schimper* 1456 (K,
   isotype).
Sorghum confine (Hochst. ex A. Rich.) Kuntze, Rev. Gen. Pl. 2: 791 (1891).

38a. var. **confinis**

Annual; culms 1–2·5 m. high, glabrous, supported by stilt roots. *Leaf-sheaths*
glabrous, produced or not at the mouth into short triangular auricles adnate to

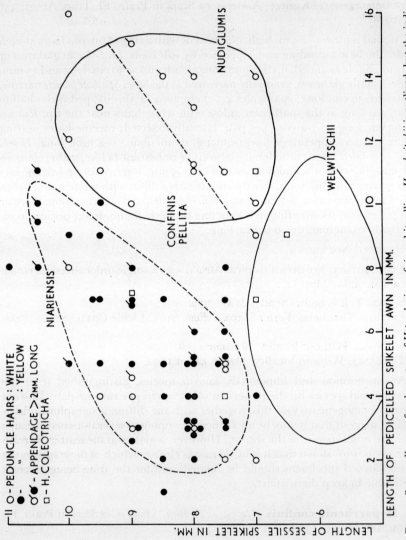

FIG. 29. Scatter diagram for characters of *Hyparrhenia confinis* and its allies. *H. welwitschii* is outlined, but the individual specimens are not plotted.

the lower part of the ligule; ligule obtuse, up to 10 mm. long; blades about 30 cm. long and up to 15 mm. wide, glabrous or shortly pilose below, narrowed at the base but not usually petiolate. *Spathate panicle* narrow, loose, coarse, about 30 cm. long; spatheoles 3–4·5 cm. long, glabrous; peduncles ½ to nearly as long as the spatheoles, white bearded above. *Racemes* 2 cm. long, 2-awned per pair, laterally exserted; raceme-bases bearing white bristles; appendage oblong, denticulate, 1–2 mm. long. *Homogamous spikelets* 9–13 mm. long, glabrous or hispidulous, the margin ciliolate. *Sessile spikelets* narrowly oblong, 8—10 mm. long; callus 1·5–3 mm. long; lower glume flat or slightly depressed, glabrous or with a few short sparse hairs; awn 4–9 cm. long; caryopsis 3–4 mm. long. *Pedicelled spikelets* 11–14 mm. long, glabrous, with a terminal bristle 10–17 mm. long.

HABITAT. Not known.

ETHIOPIA. Without locality, *Schimper* 1456, *Quartin Dillon & Petit* 70.

38b. var. **pellita** (*Hack.*) *Stapf* in Prain, Fl. Trop. Afr. 9: 354 (1918).

Andropogon confinis var. *pellitus* Hack. in DC., Monogr. Phan. 6: 642 (1889). Type: Ethiopia, *Schweinfurth* 1034 (K, isotype).

Similar to var. *confinis*, but the pedicels and the lower glume of the sessile spikelet densely silky villous with white hairs.

HABITAT. A species of dark alluvial clays.

DISTRIBUTION. Endemic to the southern Sudan (Map 11, p. 136).

SUDAN. Tozi (Sept.), *Lea* 190; Gedaref (Sept.), *Beshir* 127, 150. ETHIOPIA. Matamma, *Schweinfurth* 1034.

38c. var. **nudiglumis** (*Hack.*) *W. D. Clayton*, comb. nov.

Andropogon confinis var. *nudiglumis* Hack. in DC., Monogr. Phan. 6: 641 (1889). Type: Ethiopia, *Schweinfurth* 1043 (K, isotype). *Hyparrhenia petiolata* Stapf in Prain, Fl. Trop. Afr. 9: 352 (1918). Type: Ethiopia, *Schimper* 748 (K, holotype).

Culms 1–1·5 m. high. *Leaf-sheaths* auriculate; ligule 5–20 mm. long; leaf-blades glabrous, narrowed at the base to the subterete midrib and, particularly those from the middle part of the culm, often forming a false petiole up to 8 cm. long. *Spathate panicle* with peduncles about 1 cm. longer than the spatheoles, terminally exserted. *Raceme-base* with a linear-oblong appendage 3–4 mm. long. *Homogamous spikelets* 6·5–9 mm. long. *Sessile spikelets* 7–8·5 mm. long; lower glume pubescent or glabrous; awn 6–8 cm. long. *Pedicelled spikelets* 6–10 mm. long, the bristle 9–14 mm. long. Otherwise similar to var. *confinis*.

HABITAT. A grass of heavy black clays on alluvial flats.

DISTRIBUTION. Confined to the southern Sudan and its borders (Map 11, p. 136).

SUDAN. Abu Mutarig (Sept.), *M. N. Harrison* 397; Dilling (Oct.), *Wickens* 656; Dilling to Kadugli (Nov.), *Andrews* 34; Wairkat el Garabin (Oct.), *Beshir* 62; Jongol's Post (Sept.), *Sherif* A4003. ETHIOPIA. Matamma, *Schweinfurth* 1043; Atsegua, *Schimper* 748.

H. confinis belongs to a cluster of closely related species which have commonly been confused. It is separated by a constellation of small characters which should be considered together (Fig. 29, p. 138). One of the most useful characters is the awn of the pedicelled spikelet. It shows some overlap with other species, but fortunately this is not of much significance except in the case of *H. welwitschii*, a species easily distinguished by its 3-awned raceme-pairs.

The silky indumentum of var. *pellita* is most conspicuous, but in other respects it closely resembles the type variety and the difference is probably of little consequence.

Var. *nudiglumis* is generally a more slender plant, and the leaves have a pronounced tendency to develop false petioles, but this character is by no means confined to the variety. The tendency for the lengths of spikelets and appendages to segregate together in inverse proportion suggests that a taxonomic distinction is involved; but the range of variation in these characters is continuous and, considering also the distinctive habitat shared by the varieties, there is nothing to support a separation at species level. When the species as a whole is better known it may be found that even varietal rank overstates the degree of distinction.

39. **Hyparrhenia niariensis** *(Franch.)* *W. D. Clayton*, comb. nov.

Andropogon niariensis Franch. in Bull. Soc. Hist. Nat. Autun. 8: 330 (1895). Type: Republic of Congo, *Thollon* 1037 (P, holotype; K, isotype).

A. viancinii Franch., *l. c.*: 331 (1895). Type: Central African Republic, *Viancin* (P, holotype; K, isotype).

Cymbopogon welwitschii var. *minor* Rendle, Cat. Afr. Pl. Welw. 2: 159 (1899). Type: Angola, *Welwitsch* 2820, 7420 (both K, isosyntypes).

Andropogon nsoki Vanderyst in Bull. Agric. Congo Belge 9: 242 (1918), *nom. prov.*; 'A. usoki' Vanderyst in Bull. Soc. Bot. Belge 55: 44 (1923), *in syn sub Hyparrhenia confinis.*

A. nsoki var. *van-houttei* Vanderyst in Bull. Agric. Congo Belge 9: 242 (1918), *nom. prov.*; Vanderyst in Bull. Soc. Bot. Belge 55: 44 (1923), *in syn. sub Hyparrhenia welwitschii.*

A. nsoki var. *violascens* Vanderyst in Bull. Agric. Congo Belge 9: 242 (1918), *nom. prov.*; Vanderyst in Bull. Soc. Bot. Belge 55: 44 (1923), *in syn. sub Hyparrhenia confinis.*

39a. var. **niariensis**

Annual; culms 1–2 m. high, glabrous, supported by stilt roots. *Leaf-sheaths* glabrous, or pubescent along the margins; ligule truncate, 2–6 mm. long, laterally adnate to short auricles from the mouth of the sheath; blades up to 60 cm. long and 15 mm. wide, glabrous, or sometimes pubescent beneath, narrowed to the midrib at the base and sometimes falsely petiolate. *Spathate panicle* loose, leafy, 30–50 cm. long; spatheoles 3–5 cm. long, glabrous or with a few scattered hairs; peduncles about as long as spatheoles, bearded above with yellow hairs. *Racemes* 1·5–1·8 cm. long, 2- or sometimes 3-awned per pair, terminally exserted; raceme-bases bearded with fulvous bristles; appendage oblong to narrowly oblong, 0·5–4 mm. long. *Homogamous spikelets* 9–11 mm. long, glabrous. *Sessile spikelets* lanceolate-oblong, 7·5–11 mm. long, more or less pubescent; callus 2–3 mm. long; lower glume slightly convex or more often shallowly hollowed on the back; awn 6·5–8·5(–10·5) cm. long; caryopsis 4 mm.

long. *Pedicelled spikelets* 8–13 mm. long, terminating in a bristle 2–10 mm. long; pedicel tooth obscure.

MINOR VARIATIONS. The peduncle hairs are sometimes white. Such specimens tend to have long (3–4 mm.) raceme-base appendages and fairly short pedicelled spikelet awns, and are thus clearly distinct from *H. confinis* (Fig. 29, p. 138).

CHROMOSOME NUMBER. Congo Republic, 2n = 18 (Celarier & Harlan, 1957), as *H. confinis*.

HABITAT. In savanna and in old fallow, usually on sandy soils.

DISTRIBUTION. Around the perimeter of the Congo basin (Map 11, p. 136).

CAMEROUN. Wum (Nov.), *Brunt* 882; Nanga to Eboko (Oct.), *Vaillant* 2291; Dankali (Oct.), *Letouzey* 6098; Tibati to Foumban (Oct.), *Koechlin* 7487; Meiganga (Oct.), *Koechlin* 7142.

CENTRAL AFRICAN REPUBLIC. Krebédgé, *Chevalier* 5743; Fort de Possel, *Chevalier* 5313; Upper Oubangui, *Viancin*; Yomoko (Sept.), *Boudet* 1896; Zémio to Rafaï (Aug.), *Clair* 8; Yaloké (Oct.), *Koechlin* 3030.

REPUBLIC OF CONGO. Niari, *Thollon* 1037.

CONGO REPUBLIC. Ango (Nov.), *Germain* 4351; Basape (Dec.), *Germain* 4493; Bambili to Imadi (June), *Claessens* 817; Lodja (July), *Maillon* 14; Mvuazi (May), *Devred* 215; Kisantu (May), *Callens* 3591, 3593; Mutambo to Mukulu (June), *Quarré* 2504; Lubariku (May), *Léonard* 4442.

TANZANIA. Kigoma (Apr.), *Siwezi* 134; Kigoma, Kakombe valley (Dec.), *Pirozynski* P96.

ZAMBIA. Mulundu (Apr.), *Symoens* 8460.

ANGOLA. Pungo Andongo, *Welwitsch* 2820; Maludi (Nov.), *Gossweiler* 13854; Camaquezo (June), *Gossweiler* 14080.

39b. var. **macrarrhena** *(Hack.) W. D. Clayton*, comb. nov.

Andropogon confinis var. *macrarrhenus* Hack. in DC., Monogr. Phan. 6: 642 (1889). Type: Sudan, *Schweinfurth* 2618 (K, isotype).
Hyparrhenia macrarrhena (Hack.) Stapf in Prain, Fl. Trop. Afr. 9: 355 (1918).

Racemes 2-awned; raceme-base with an obscure scarious rim up to 0·2 mm. long. *Sessile spikelets* with the lower glume firmly chartaceous and hollowed into a U-shape along the back. Otherwise similar to the preceding.

MINOR VARIATIONS. *Le Testu* 3320 from Central African Republic has yellow peduncle hairs and no appendage, but there are 4 awns per raceme-pair and a homogamous pair at the base of both racemes. Although obviously anomalous, I hesitate to propose a new taxon until the group as a whole is better understood.

HABITAT. Savanna woodland.

DISTRIBUTION. Endemic in the southern Sudan.

SUDAN. Aluackluack (Oct.), *M. N. Harrison* 1412; Bongoland, *Schweinfurth* 2618 (or 2018, writing illegible).

H. niariensis has commonly been confused with *H. confinis*, from which it is best separated by its shorter-awned pedicelled spikelets and generally yellow peduncle beard (Fig. 29, p. 138). The two species are contiguous, and the decision to separate them is reinforced by their distinct geographical distribution, and their very different ecological habitats. *H. confinis* grows on black clays, while *H. niariensis* is a species of the normal more or less sandy savanna soils.

Three awns per raceme-pair are sometimes found, but they are mixed with 2-awned pairs in the same panicle. This feature, together with the larger sessile spikelets, helps to separate such specimens from the strictly 3-awned *H. welwitschii*. Variation in these two species however is also contiguous, and intermediates, such as the type of *H. welwitschii* var. *minor*, are not easy to place.

Var. *niariensis* includes specimens with a very short appendage and distinctly hollowed lower glume of the sessile spikelet. *H. macrarrhena* represents the extreme expression of this tendency, and I have accordingly reduced it to a variety; present evidence is insufficient to advocate the complete union of the taxa. It may be observed in passing that the broadly concave back of the lower glume bears no resemblance to the narrow median groove of *Hyperthelia*.

40. **Hyparrhenia welwitschii** *(Rendle) Stapf* in Prain, Fl. Trop. Afr. 9: 356 (1918).

Cymbopogon welwitschii Rendle, Cat. Afr. Pl. Welw. 2: 157 (1899). Types: Angola, *Welwitsch* 2955, 2956, 3000 (K, isosyntype), 7190 (K, isosyntype), 7248 (K, isosyntype).

Andropogon chrysopogon Welw. ex Rendle, *l. c.* (1899), *in synon.*

A. welwitschii (Rendle) K. Schum. in Just, Bot. Jahresber. 27, 1: 454 (1901).

Hyparrhenia gracilescens Stapf in Prain, Fl. Trop. Afr. 9: 357 (1918). Type: Guinea, *Chevalier* (K, syntype); Nigeria, *Dalziel* 292 (K, syntype).

Annual; culms 0·2–3 m. high, glabrous or with a ring of brown hairs at each node, coarsely tufted, often supported by stilt roots. *Leaf-sheaths* glabrous; ligule truncate, about 3 mm. long; blades up to 60 cm. long and 12 mm. wide, glaucescent, glabrous, narrowed at the base. *Spathate panicle* large, loose, leafy, 30–60 cm. long, brown bearded at the nodes; spatheoles 3–5 cm. long, glabrous; peduncles a little shorter than the spatheoles, pilose above with white to yellow hairs. *Racemes* 1·2–1·7 cm. long, 3-awned per pair, laterally exserted; raceme-bases bearing pale rufous bristles; appendage oblong, irregularly lobed, reddish, 0·5–1 mm. long. *Homogamous spikelets* 7–10 mm. long, glabrous. *Sessile spikelets* linear-oblong, 5–7 mm. long; callus 1·5 mm. long; lower glume greyish, glabrescent or sparsely pubescent with white hairs; awn 4–7 cm. long; caryopsis 3 mm. long. *Pedicelled spikelets* 6–8 mm. long, glabrous, terminating in a fine bristle 2–11 (rarely 14) mm. long; pedicel tooth short, broadly triangular.

CHROMOSOME NUMBER. Katanga-Zambia region, 2n = 40, precise locality uncertain!.

HABITAT. A grass of light to moderate shade in savanna woodland (*e.g.*, *Daniellia* or *Brachystegia* types), favouring deep soils with adequate moisture. Commonly found in gregarious patches.

DISTRIBUTION. Western Africa from Guinea to Angola, but reaching eastwards to the Comoro Is. (Map 12).

GUINEA. Fouta Djallon, *Chevalier* 20196; Labé (Nov.), *Adames* 409; Kindia, *Jacques-Félix* 231; Bafing (Oct.), *Adam* 12658; Timbo (Oct.), *Pobéguin* 1797.

Sierra Leone. Falaba (Nov.), *Miszewski* 17, *J. K. Morton* 2857; Mamodia (Nov.), *Glanville* 330; Musaia (Feb., Dec.), *Deighton* 4279, 4433.

Ivory Coast. Touba (Oct.), *Aké-Assi* 8210; Madinani (Oct.), *Boudet* 3314; Kouilly (Dec.), *Portères* 424; Séguéla (Oct.), *Adjanohoun* 286a.

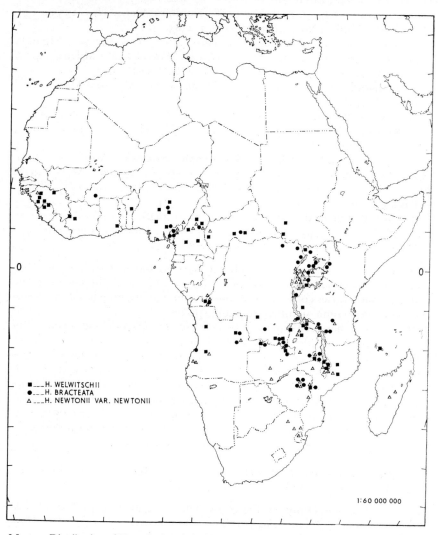

H. WELWITSCHII
H. BRACTEATA
H. NEWTONII VAR. NEWTONII

1:60 000 000

Map 12. Distribution of *Hyparrhenia welwitschii*, *H. bracteata* and *H. newtonii* var. *newtonii*

Ghana. Biakpa (Nov.), *J. K. Morton*.

Dahomey. Parakou (Nov.), *Risopoulos* 1296.

Nigeria. Anara F.R., Kan Gimi (Oct.), *Keay* FHI 5438; Zungeru (Nov.), *Freeman* S189; Jemaa (Nov.), *Clayton* 1434; Lokoja (Oct.), *Dalziel* 292; Ogoja Prov., *Roseveare* 20/30a.

Cameroun. Ngaundéré (Oct.), *Breteler* 528; Djouroum (Dec.), *Raynal* 12124b; Meiganga to Ngaundéré (Oct.), *Koechlin* 7180, 7219; Yaundé, *Zenker & Staudt* 98; Bertoua (Dec.), *Baldwin* 13952.

CENTRAL AFRICAN REPUBLIC. Bouar (Sept., Dec.), *Boudet* 1555, *Koechlin* 6284; Gougali (Oct.), *Boudet* 2255; Ouaka (Feb., Oct.), *Tisserant, Le Testu* 2699.

CONGO REPUBLIC. Kisantu to Madimba (Apr.), *Compère* 1976; Mvuazi (June), *Devred* 552; Elisabethville (May), *Symoens* 10985, 11009.

SUDAN. Wau (Oct.), *J. K. Jackson* 3988; Tambura (Oct.), *Myers* 13530.

UGANDA. Yumbe (Nov.), *Langdale-Brown* 2383.

TANZANIA. Biharamulo (May), *Tanganyika Forest Herbarium* 412; Mpanda (Apr.), *Boaler* 266; Kasanga (Mar.), *McCallum Webster* T226; Milepa (Apr.), *Vesey-FitzGerald* 1742; Rungwe Distr. (Apr.), *R. M. Davies* 522.

MOZAMBIQUE. Malema (Mar.), *Torre & Paiva* 11273; Mocuba (Mar., June), *Torre* 4898, 5457*b*.

MALAWI. Fort Manning (Apr.), *G. Jackson* 767; Salima to Citala (Apr.), *G. Jackson* 474; Namweras highlands (Apr.), *Lawrence* 381; Zomba (May), *Cormack* 465, *Wiehe* N102.

ZAMBIA. Kawambwa (Apr.), *Angus* 2758; Ndundu (Apr.), *McCallum Webster* A271; Kasama (Apr.), *Phipps & Vesey-FitzGerald* 2998; Mwinilunga, Matonchi Farm (Oct.), *Milne-Redhead* 2759; Ndola (Apr.), *G. Jackson* 6.

RHODESIA. Nyumkwarara, *Gilliland* 2022.

ANGOLA. Golungo Alto, *Welwitsch* 3000, 7190, 7248; Henrique de Carvalho (Apr.), *Gossweiler* 11540; Huambo, Namba (Apr.), *Barbosa & Correia* 9150.

COMORO Is. Maouéni (May), *Bosser* 18318.

A species distinguished by its small spikelets and 3-awned raceme pairs, but with some resemblance to *H. niariensis* (Fig. 29, p. 138). The colour of the peduncle hairs varies considerably from off-white to bright yellow, and seems to be of no taxonomic significance in this species.

H. gracilescens has been separated on account of its glabrous culm nodes, but I can find no other difference between these two species, and the presence or absence of hairy nodes seems to be uncorrelated with any other character. It is not even a clear-cut character, for some specimens, including one of the syntypes of *H. gracilescens*, have bearded panicle nodes. The bearded culm nodes can be a most striking sight when the sun glints on the brown hairs to form diaphanous halos of golden light at intervals up the culm. No doubt this explains the taxonomic importance accorded to the character, an importance that is not confirmed by a more dispassionate assessment.

41. **Hyparrhenia bracteata** (*Humb. & Bonpl. ex Willd.*) *Stapf* in Prain, Fl Trop. Afr. 9: 360 (1918).

Andropogon bracteatus Humb. & Bonpl. ex Willd., Sp. Pl. 4: 914 (1806); Hack. in DC., Monogr. Phan. 6: 643 (1889). Type: Venezuela, *Humboldt* in Herb. Willd. 18655 (B, holotype).

Cymbopogon humboldtii Spreng., Pugill. 2: 15 (1815), *nom. superfl.* Based on *Andropogon bracteatus.*

Anthistiria foliosa H.B.K., Nov. Gen. Sp. 1: 191 (1816). Type: Venezuela (not seen).

A. reflexa H.B.K., *l. c.* (1816), *nom. superfl.*

Cymbopogon foliosus (H.B.K.) Roem. & Schult., Syst. Veg. 2: 835 (1817).

C. reflexus Roem. & Schult., *l. c.*: 834 (1817), *nom. superfl.*

Anthistiria humboldtii Nees, Agrost. Bras.: 369 (1829), *nom. superfl.*

A. pilosa J. S. & C. B. Presl, Rel. Haenk. 1: 348 (1830). Type: Peru (not seen).

Andropogon trachypus Trin. in Mem. Acad. Sci. Petersb., sér. 6, 2: 280 (1832). Type: Brazil (not seen).

A. lindenii Steud., Syn. Pl. Glum. 1: 389 (1854). Type: Colombia, *Linden* 1556 (K, isotype).

Anthistiria andropogonoïdes Steud., Syn. Pl. Glum. 1: 402 (1854). Type: Venezuela, *Funck* 743 (K, isotype).

Hyparrhenia foliosa (H.B.K.) Fourn., Mex. Pl. 2: 67 (1886).

Themeda foliosa (H.B.K.) Balansa in Journ. de Bot. 4: 116 (1890).

Sorghum bracteatum (Humb. & Bonpl. ex Willd.) Kuntze, Rev. Gen. Pl. 2: 791 (1891).

Andropogon setifer Pilger in Wiss. Ergebn. Deutsch. Zentr. Afr. Exped. 1907–8, 2: 44 (1910). Type: Tanzania, *Mildbraed* 98 (K, isotype).

Cymbopogon bracteatus (Humb. & Bonpl. ex Willd.) Hitchc. in Contrib. U.S. Nat. Herb. 17: 209 (1913).

C. setifer (Pilger) Pilger in Engl., Bot. Jahrb. 54: 287 (1917).

Andropogon nlemfuensis Vanderyst in Bull. Agric. Congo Belge 9: 242 (1918), *nom. prov.*; & in Bull. Soc. Bot. Belge 55: 44 (1923), *in synon.*, sub *Hyparrhenia bracteata*. (All but one of the original sheets were *H. newtonii*.)

Cymbopogon pilosovaginatus De Wild. in Bull. Jard. Bot. Brux. 6: 17 (1919). Type: Congo Republic, *Ringoet* 11, *Bequaert* 444, 987 (all BR, syntypes).

Andropogon pilosovaginatus De Wild., *l. c.*, (1919), *in synon.*

Hyparrhenia contracta Robyns, Fl. Agrost. Congo Belge 1: 189 (1929); & in Bull. Jard. Bot. Brux. 8: 241 (1930). Type: Congo Republic, *Quarré* 273 (K, isotype).

Perennial; culms 0·6–2·5 m. high, densely tufted, glabrous or sometimes villous for a short distance below the nodes. *Leaf-sheaths* usually sparsely hirsute, sometimes densely villous, rarely quite glabrous, the basal sheaths nearly always tomentose with pale brown hairs below; ligule truncate, 1–2 mm. long; blades up to 60 cm. long and 4 mm. wide, rigid, glaucescent, scabrid along the margins, nearly always pubescent below. *Spathate panicle* narrow, dense, 20–60 cm. long, the primary tiers with 3–4 compound rays, these bearing secondary and tertiary tiers; spatheoles 2–3 cm. long, commonly appressedly hirsute but sometimes hairy only along the margins; peduncles a little shorter than the spatheoles, copiously bearded with stiff yellow hairs near the tip. *Racemes* 0·5–1·5 cm. long, 2–4-awned per pair, dark purple, laterally exserted; raceme-bases slightly unequal, the upper 1·5–2 mm. long, flattened or subterete, bearded with stiff yellow hairs; appendage narrowly oblong, irregularly bifid or lobed, purple, 1–2·5 mm. long, glabrous, pubescent or even villous. *Homogamous spikelets* 4–7 mm. long, glabrous with scabrid margins. *Sessile spikelets* narrowly linear-oblong, 4–6 mm. long, glabrous; callus acute, 1 mm. long; lower glume coriaceous, glabrous, the middle nerve raised between two fine grooves; awn 1–2·5 cm. long; caryopsis 2 mm. long. *Pedicelled spikelets* 4–6(–7) mm. long, glabrous, muticous or with a short mucro seldom over 1 mm. long; pedicel tooth obtusely triangular, obscure.

ILLUSTRATIONS. Mart., Fl. Bras. 2, 3: t. 64 (1883); Pilger in Engl. & Prantl, Nat. Pflanzenfam. 14e: 176, t. 94 (1940).

MINOR VARIATIONS. Odd specimens with the pedicel tooth up to 2 mm. long are found in the region of overlap between *H. bracteata* and *H. newtonii* (Fig. 30, p. 146); they probably result from introgression, and their affinities can

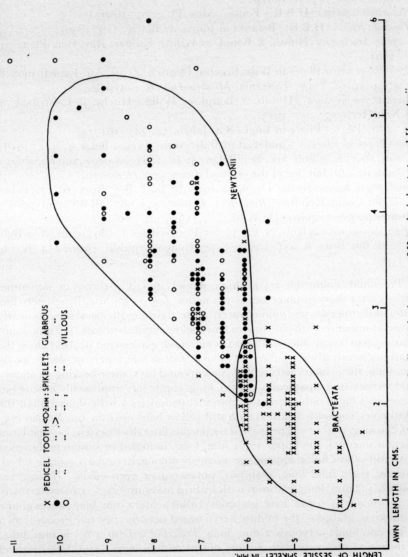

FIG. 30. Scatter diagram for diagnostic characters of *Hyparrhenia bracteata* and *H. newtonii*.

only be decided by consideration of the characters as a whole. The number of awns per raceme-pair seems to be of little significance in this species since it varies even within the same panicle. Specimens with the smallest spikelets seem to be more prevalent towards the southern limit of the range. I have seen two specimens (Tanzania, *Geilinger* 2062; and Zambia, Kawimbe, *McCallum Webster* A247) with villous sessile spikelets.

A much more puzzling deviation is represented by the following South American specimens: Mexico, Mirador, *Linden* 104, *Liebmann* 46; Colombia, *Purdie*, *Little* 8560. They have sessile spikelets, 7–9 mm. long, and awns 3·3–5·1 cm. long, but display the dense panicle and toothless pedicel tip of *H. bracteata*. I am inclined to regard them as mere variants of the latter, for typical *H. newtonii* has not been found in America. Moreover occasional exuberant awn lengths seem to be a feature of the group; they occur in *H. newtonii*, and I have seen one such specimen (Zambia, *Mitchell* 25/44) among the African material of *H. bracteata*.

Habitat. Confined to shallow drainage lines (dambos or mbugas) and similar marshy places.

Distribution. The western half of the African continent from Ghana to Rhodesia (Map 12, p. 143); also in South America from Mexico to Paraguay.

Upper Volta. Bobo-Dioulasso (June), *Adam* 15093.

Ivory Coast. Without locality, *Bégué* 136.

Nigeria. Vodni, *Saunders* 9; Obudu plateau (Nov.), *Tuley* 1023.

Cameroun. Nyen (May), *Brunt* 1132; Wum (Nov.), *Brunt* 868; Bambui (July), *Bumpus* Bam/26; Bamenda (Dec.), *Baldwin* 13833; Bamenda to Bali (Apr.), *Brunt* 1667; Mankim (Nov.), *Raynal* 11930; Meiganga to Ngaundéré (Oct.), *Koechlin* 7199; Manengouba (Dec.), *Hédin*.

Central African Republic. Bagondo (Oct.), *Boudet* 3213; Ippy (Dec.), *Descoings* 11716.

Congo Republic. Djugu (Sept.), *Froment* 530; Kisantu (Apr.), *Germain* 2134; Marungu (Apr.), *Dubois* 1324; Kapanda valley (Dec.), *Homblé* 987; Tumbwe (Mar.), *Symoens* 8384; Elisabethville (May), *Bequaert* 444; Kafubu (Apr.), *Quarré* 273; Shinsenda (May), *Ringoet* 11.

Burundi. Bururi (June), *Taton* 1039; Luvironza valley (Mar.), *Germain* 8316.

Uganda. Gulu (Nov.), *A. S. Thomas* 4007; Wabusana to Luwero (July), *Langdale-Brown* 2251; Kampala (Mar.), *Michelmore* 1280; Masaka, *Maitland* 762; Katera (June), *A. S. Thomas* 1280.

Kenya. Kitale (June), *Bogdan* 3437; Bungoma (June), *Bogdan* 4059.

Tanzania. Itara, *Mildbraed* 98; Bukoba (Feb.), *Haarer* 2505; Njombe, *Emson* 128; Lupembe, Msima, *Schlieben* 1009; Rungwe (May), *Stolz* 1301.

Mozambique. Garuso (Apr.), *Gilliland* 1829.

Malawi. Kota-kota (July), *Brass* 16975; Dzalanyana (Dec.), *G. Jackson* 696; Dedza (Mar.), *Wiehe* N/464; Mvai (June), *Wiehe* N/589; Zomba (Nov.), *Wiehe* N/365.

Zambia. Abercorn (May), *Richards* 9932; Mwinilunga (Nov.), *Milne-Redhead* 3120; Luanshya (Apr.), *Angus* 2847; Fort Jameson (May), *van Rensburg* 2116; Kalomo, Ibula (Mar.), *Trapnell* 1492.

Rhodesia. Banket (Mar.), *Rattray* in GH 15780; Salisbury (Mar., Apr.), *Eyles* 4769, *Brain* 3531; Hartley (June), *R. M. Davies* in GH 32697; Macheke (May), *Corby* in GH 20978.

ANGOLA. Lunda, Dala (Apr.), *Gossweiler* 11546; Henrique de Carvalho (Apr.), *Gossweiler* 11608; Benguela, *Gossweiler* 2577, 3124, 3925.

MEXICO. Orizaba (Oct.), *Bourgeau* 3270.

GUATEMALA. Quiriguá to Los Amates (Mar.), *Weatherwax* 101; Cuajiniquilapa, *Heyde & Lux* 6279.

GUYANA. Rupununi, *Melville*.

BRAZIL. Without locality, *Riedel* 156, *Burchell* 4457; Goias, *Glaziou* 22378; Minas Gerais, *Widgren*; Rio de Janeiro, *Glaziou* 13336; Paraná, Jaguariahyou (Apr.), *Dusén* 11616.

PARAGUAY. Upper reaches of Rio Apa (Mar., Apr.), *Hassler* 11057, 11096; Rio Apa to Rio Aquidaban (Jan.), *Fiebrig* 5088.

PANAMA. Boquete (Feb.), *Seemann* 1562.

COLOMBIA. Santa Marta (July), *H. H. Smith* 2121; San Antonio (Nov.), *Langlasse* 36; without localities, *Linden* 1556, *Triana* 336, *André* 496.

VENEZUELA. Without locality, *Funcke* 743; Caracas, *Fendler* 1657.

BOLIVIA. Rurrenabaque, *O. E. White* 1125.

H. bracteata is very closely related to *H. newtonii*, sharing with that species the characteristic purple racemes framed in yellow hairs from the peduncle and raceme-bases. It is distinguished by its dense narrow panicles, smaller spikelets, shorter awns and muticous pedicelled spikelets, but these characters intergrade with *H. newtonii* (Fig. 30, p. 146). More definite characters are provided by the obsolete pedicel tooth and the preference for a swampy habitat, so that its segregation as a distinct species is fully justified. There is almost certainly some introgressive hybridization along the ecotone between swampy and dry soils, and in a few cases it is almost impossible to distinguish between the two species. It may be noted that specimens of *H. newtonii* with small spikelets nearly always have the awn of the pedicelled spikelet 2 mm. long or more, a point which is often helpful in separating the species; but none of the distinguishing characters are reliable on their own, and they must be considered in conjunction.

42. **Hyparrhenia newtonii** (*Hack.*) *Stapf* in Prain, Fl. Trop. Afr. 9: 363 (1918).

Andropogon newtonii Hack. in Bol. Soc. Brot. 3: 137 (1885); Hack. in DC., Monogr. Phan. 6: 644 (1889). Type: Angola, *Newton* (K, isotype).

Sorghum newtonii (Hack.) Kuntze, Rev. Gen. Pl. 2: 792 (1891).

Andropogon lecomtei Franch. in Bull. Soc. Hist. Nat. Autun 8: 329 (1895). Type: Republic of Congo, *Lecomte* (P, holotype).

Cymbopogon lecomtei (Franch.) Rendle in Journ. Linn. Soc., Bot. 40: 227 (1911).

Hyparrhenia cirrulosa Stapf in Prain, Fl. Trop. Afr. 9: 365 (1918). Type: Congo Republic, *Rogers* 10399 (K, holotype).

H. lecomtei (Franch.) Stapf, *l. c.*: 362 (1918).

H. stolzii Stapf, *l. c.*: 364 (1918), non *Cymbopogon stolzii* Pilger (1917). Type: Tanzania, *Stolz* 2629 (K, holotype).

Andropogon bisulcatus Chiov. in Nuov. Giorn. Bot. Ital. N.S., 26: 60 (1919), *in synon.*

Hyparrhenia bisulcata Chiov., *l. c.* (1919). Types: Congo Republic, *Bovone* 88, 114 (both TO, syntypes).

Andropogon nlemfuensis var. *villosus* Vanderyst in Bull. Soc. Bot. Belge 55: 44 (1923), *in synon.*, sub *Hyparrhenia lecomtei*.

Hyparrhenia lecomtei var. *bisulcata* (Chiov.) Robyns, Fl. Agrost. Congo Belge 1:
192 (1929); & in Bull. Jard. Bot. Brux. 8: 242 (1930).

H. squarrulosa Peter in Fedde, Rep. Sp. Nov. Beih. 40 (1): 375 & Anh. 118
(1936). Type: Tanzania, *Peter* 38429 (B, destroyed during second world
war).

42a. var. **newtonii**

Perennial; culms 60–120 cm. high, densely tufted, glabrous. *Leaf-sheaths*
glabrous to thinly hirsute, the basal sheaths tomentose or glabrous below;
ligule truncate, about 1 mm. long; blades up to 30 cm. long and 3 mm. wide,
rigid, light green to glaucescent, scabrid along the margins, glabrous or
pubescent beneath. *Spathate panicle* diffuse or scanty, 15–30 cm. long, the
primary tiers simple or with 1–3 compound rays; spatheoles 2·5–5 cm. long,
glabrous or villous along the margin, rarely villous all over; peduncles a little
shorter than the spatheole, pilose with stiff yellowish hairs towards the top.
Racemes 1·5–2 cm. long, 2–4-awned per pair, dark purple, laterally exserted;
raceme-bases unequal, the upper 1·5–3(–4) mm. long, flattened below, subterete above, bearded with stiff yellow, or sometimes pallid, hairs; appendage
linear, entire or bidentate, purple, 1–3 mm. long. *Homogamous spikelets* 5–10
mm. long, glabrous. *Sessile spikelets* linear-oblong, 6–10 mm. long, glabrous;
callus acute to pungent, 1·5–2 mm. long; lower glume coriaceous, glabrous or
sometimes hairy near the tip, the middle nerve raised between two fine grooves;
awn 2·2–5·5 cm. long; caryopsis 2·5–3·5 mm. long. *Pedicelled spikelets* 5–10 mm.
long, glabrous, terminating in a bristle 1–5 mm. long; pedicel tooth subulate,
0·2–1·5 mm. long.

MINOR VARIATIONS. Two specimens from the Congo (Tansi, *Callens* 2484;
Bwana Mutombo, *Callens* 3042) have sessile spikelets 10–12 mm. long and
awns 7–7·3 cm. long. Although lying outside the normal range of *H. newtonii*,
it would be most unsafe to conclude that these specimens constitute a separate
taxon.

G. Jackson 819 from Mposa, Malawi is quite atypical. Raceme-bases sparsely
hairy, not deflexed; appendage 4 mm. long; spikelet size and colouring as for
H. newtonii, but grooves of lower glume coalescent into a single deep median
groove. It is either an exceptional form of *H. newtonii*, or a product of introgression from *Hyperthelia dissoluta* (Nees) W. D. Clayton.

CHROMOSOME NUMBER. Zambia, 2n = 40 (Moffett & Hurcombe, 1949).

HABITAT. In the northern part of its range (Cameroun, Congo, Uganda,
Tanzania) it occurs mainly on lithosols and stony hillsides. To the south
(Zambia, Rhodesia and Malawi) it is found mainly in dappled shade beneath
Brachystegia woodland. This duality of habitat may be a matter of compensat-
ing edaphic factors in the more humid climatic regime to the north; or it may
merely reflect the relative frequency of these habitats on the Miocene pediplain
of Rhodesia, and in the generally more dissected country to the north. Some
collectors have obviously failed to distinguish satisfactorily between *H. newtonii* and *H. bracteata* and have described a composite habitat.

DISTRIBUTION. Western Africa from Guinea to Angola, Rhodesia and
Transvaal (Map 12, p. 143). Also in Madagascar, Thailand, Indo-China and
Indonesia.

THAILAND. Chiengmai, *Smitinand* 3763; Phu Krading (Nov.), *Smitinand* 4982, 9540, *Larsen* 6264.

VIETNAM. Mt. Bavi (Dec.), *Balansa* 1728; Ouanbi (Oct.), *Balansa* 384; Darlet (Feb.), *Smitinand* 6397.

INDONESIA. Soemba (May), *Monod de Froideville* 1737.

GUINEA. Labé (Nov.), *Adames* 411.

SIERRA LEONE. Loma Mts. (Nov.), *J. K. Morton* SL 2744.

NIGERIA. Vogel Peak, Dawo (Nov.), *Hepper* 1424.

CAMEROUN. Bambui (Dec.), *Pedder* 7; Bamenda (June), *Brunt* 1178, 1179; Banso to Bamenda (Oct.), *Tamajong* FHI 23486; Ngaundéré to Tibati (Oct.), *Koechlin* 7467; Ngaundéré (Oct.), *Koechlin* 7233; Meiganga (Oct.), *Koechlin* 7132.

CENTRAL AFRICAN REPUBLIC. Bouar (Sept.), *Koechlin* 6341; Yalinga (Oct.), *Le Testu* 3319.

REPUBLIC OF CONGO. Kitabi (Nov.), *Lecomte*; Brazzaville (May), *Koechlin* 2670.

CONGO REPUBLIC. Kasindi to Lubango (Jan.), *Lebrun* 4777; Kisantu (May), *Callens* 3608; Inkissi valley, *Vanderyst* 414; Sakania to Ndola (Apr.), *Symoens* 9435; Tschinsenda, *F. A. Rogers* 10399.

RWANDA. Katitumba (Jan., May), *Troupin* 3220, 5921; Nyakayaga, *Lebrun* 9457; Kibungu, Tuntu (Jan.), *Troupin* 5861.

BURUNDI. Karuzi (Jan.), *Van der Ben* 2424.

UGANDA. Mabungo (Oct.), *Snowden* 1496; Rukungirc (Apr.), *A. S. Thomas* 4434; Gayaza (July), *Snowden* 1401; Kabula (Mar.), *Michelmore* 1318b; Sese Is., Kalangala (Feb.), *Greenway & Thomas* 7167.

TANZANIA. Bukoba, Minziro F.R. (Jan.), *Procter* 1126; Keza (May), *Watkins* 422; Sumbawanga (Apr.), *Phipps & Vesey-FitzGerald* 3320; Mbeya, *Jacobsen* 8; Utengule (Apr.), *Stolz* 2629.

MOZAMBIQUE. Serra da Pandalanjala (May), *Mendonça* 4254.

MALAWI. Fort Manning (Apr.), *G. Jackson* 769; Chitedzi (June), *Wiehe* N619; Zomba (May), *Whyte, G. Jackson* 833; Shire highlands, *Buchanan*.

ZAMBIA. Mporokoso (Jan.), *Richards* 12079; Abercorn (Apr.), *Vesey-Fitz-Gerald* 162b; Luwingu (May), *Astle* 620; Mufulira (May), *Cruse* 331; Kalomo (May), *Trapnell* 1446.

RHODESIA. Salisbury (May), *Blackwell* in *Eyles* 2233, *Gilliland* 14; Maran-dellas (Apr.), *Eyles* 3876; Inyanga, *Norlindh & Weimarck*; Melsetter (June), *Crook* 484.

ANGOLA. Without locality, *Welwitsch* 7514; Dala (Apr.), *Gossweiler* 11226; Nova Lisboa (May), *Gossweiler* 11303; Munhango (May), *Gossweiler* 11304; Humpata, *F. Newton*.

MADAGASCAR. Without locality, *Baron* 5205, *Perrier de la Bâthie* 21; Tanan-arive (May), *Bosser* 15992; Faratsiho (Feb., July), *Bosser* 10837, *Tateoka* 3555; Angavo (Mar.), *Decary* 7329.

SWAZILAND. Mabane (Jan.), *F. A. Rogers* 11701.

SOUTH AFRICA. Transvaal: Potgietersrust (Dec.), *Story* 1645; Middleburg (Dec.), *Schlechter* 4054.

42b. var. **macra** *Stapf* in Prain, Fl. Trop. Afr. 9: 364 (1918). Type: Rhodesia, *Craster* 59 (K, lectotype).

Lower glume of sessile spikelets, and sometimes also of pedicelled spikelets, pubescent to villous. Otherwise similar to var. *newtonii*.

Habitat. As for var. *newtonii*.

Distribution. Virtually restricted to the southern part of the range of var. *newtonii*.

Congo Republic. Marungu (Apr.), *Dubois* 1232, 1339; Elisabethville (Apr., Oct.), *Gathy* 754, *F. A. Rogers* 10297.

Rwanda. Kigali to Kakisumba (Jan.), *Troupin* 5798; Parc Nat. Kagera, Mashoro hill (Feb.), *Troupin* 5909.

Uganda. Rukungire (Apr.), *A. S. Thomas* 3824; Ankole, *Snowden* 1278*b*.

Tanzania. Iringa (May), *Schlieben* 1007, *Emson* 467, *Wolfe* 10; Sumbawanga (Mar.), *McCallum Webster* T136; Mmemya Mtn. (Feb.), *Bullock* 3708.

Mozambique. Pico Namuli (Apr.), *Torre* 5131.

Malawi. Nyika (June), *G. Jackson* 531; Vipya (Mar.), *G. Jackson* 435; Champoyo (Mar.), *G. Jackson* 1286; Zomba (May), *Brass* 16093; Mlanje (Nov.), *Wiehe* N342.

Zambia. Abercorn (May), *Vesey-FitzGerald* 3381; Luwingu (Apr.), *Astle* 537; Mufulira, *Eyles* 8365; Kanona (Apr.), *Phipps & Vesey-FitzGerald* 2963; Namwala (Apr.), *van Rensburg* 2062.

Rhodesia. Salisbury (Apr.), *Craster* 59, *Brain* 4249; Digglefold (Feb.), *Corby* 385; Inyanga (Mar.), *Rattray* 1393; Melsetter (Apr.), *Whellan* 1238.

Angola. Dala (Apr.), *Gossweiler* 11633.

South Africa. Transvaal: Sabie (Apr.), *Louw* 2720.

H. newtonii is customarily divided into two species, *H. lecomtei* having smaller spikelets than *H. newtonii sensu stricto*. However, as can be seen from figure 30 (p. 146), the variation in spikelet size is quite continuous. Several other characters display a definite gradient in the same direction as spikelet size; thus *H. newtonii sensu stricto* tends to have a scantier panicle, the raceme-pairs are nearly always 2-awned, and there is a higher proportion of specimens with glabrous leaf-blades or basal sheaths. Unfortunately a more careful analysis reveals that these characters are correlated only in the most general way, and they prove a most haphazard means of discriminating between the species.

Geographically *H. lecomtei* is more frequent in the northern part of the range (also in Madagascar and Thailand), while *H. newtonii sensu stricto* becomes predominant in the south. There seems to be no ecological difference between the species, for *H. lecomtei* spans both the main habitat types. I have therefore concluded that, although there is evidence of clinal variation, no satisfactory disjunction can be discerned, and only one species should be recognized.

I have retained variety *macra*, whose size range corresponds roughly with that of *H. newtonii sensu stricto*, but suspect that this distinction is not of much biological significance. The syntypes include both hairy and glabrous specimens, and it has been necessary to choose a lectotype.

V. Sect. **Arrhenopogonia** *W. D. Clayton*, sect. nov.; a sect. *Pogonopodia* Stapf spiculis homogamis ad basin racemi superioris suppetentibus, spatheolis augustioribus et pedunculis longioribus differt; a sect. *Apogonia* spiculis homogamis pectinato-ciliatis et basibus racemorum plerumque barbatis distinguenda. Type species: *H. arrhenobasis* (Hochst. ex Steud.) Stapf.

Annuals or perennials, the former sometimes very small. *Leaf-sheaths* glabrous; ligule 0·5–1 mm. long; blades glabrous. *Spathate panicle* scanty, of few raceme-pairs; spatheoles very narrow; peduncles from ½ to rather longer than the spatheole. *Racemes* usually not deflexed, densely spiculate, terminally or laterally exserted on the recurved peduncle; raceme-bases subequal, short, flattened, usually clothed in stiff fulvous glassy bristles, the tips obliquely truncate, rarely with a lobed rim up to 0·2 mm. long. *Homogamous pairs* 1 or 2 at the base of both racemes; spikelets lanceolate, acuminate, the margins stiffly pectinate-ciliate. *Sessile spikelets* narrowly lanceolate, becoming dark brown at maturity; callus acute to pungent; upper lemma lobes 0·5–0·8 mm. long; awn fulvously hairy. *Pedicelled spikelets* narrowly lanceolate, acuminate to aristulate, stiffly ciliate on the margins; callus short, oblong, truncate; pedicel tooth short, triangular.

A homogeneous group of close-knit species whose affinities pose a difficult problem. They have the flat, bearded, ebracteate raceme-bases of sect. *Pogonopodia*, although the beard may be absent in *H. multiplex* and *H. anemopaegma*. On the other hand they have two pairs of homogamous spikelets at the base of both racemes (only one pair in *H. anemopaegma* and some specimens of *H. arrhenobasis*) such as occur in sect. *Apogonia*, but the bearded raceme-bases and strongly ciliate homogamous spikelets are foreign to this section. Yet another relationship is suggested by the leathery, pallid, concave lower glume of the immature sessile spikelet, which resembles the condition found in some species of sect. *Hyparrhenia*. This mixture of inconstant characters is most confusing, and led Stapf (1918) to place the species in two different sections. They fit comfortably in neither of these, and the most rational solution seems to be the creation of a separate section.

43. **Hyparrhenia multiplex** (*Hochst. ex A. Rich.*) *Anderss. ex Stapf* in Prain, Fl. Trop. Afr. 9: 374 (1918).

Anthistiria multiplex Hochst. ex A. Rich., Tent. Fl. Abyss. 2: 449 (1851). Type: Ethiopia, *Schimper* 1637 (K, isotype).
Andropogon multiplex (Hochst. ex A. Rich.) Hack. in DC., Monogr. Phan. 6:631 (1889).
Sorghum multiplex (Hochst. ex A. Rich.) Kuntze, Rev. Gen. Pl. 2: 792 (1891).
Hyparrhenia multiplex var. *leiopoda* Stapf in Prain, Fl. Trop. Afr. 9: 375 (1918). Type: Ethiopia, *Schimper* 349 (K, holotype).

Slender erect annual; culms 5–35 cm. high. *Leaf-blades* 2–20 cm. long and up to 2 mm. wide. *Spathate panicle* of 1–4 raceme-pairs; spatheoles at first inflated, later linear to narrowly lanceolate, 3–15 cm. long, green and leaf-like becoming scarious and straw-coloured, glabrous, terminating in a short lamina up to 2 cm. long; peduncles rather variable in length, glabrous or pilose with whitish hairs near the tip. *Racemes* seldom deflexed, about 1 cm. long, 3–11-awned per pair; raceme-bases bearded or sometimes glabrous. *Homogamous pairs* 2 at the base of each raceme; spikelets 7–8 mm. long, olive green when young, pectinate-ciliate on the margins and sometimes also on the middle nerve, otherwise glabrous. *Sessile spikelets* 6–8 mm. long, hispidulous; callus 2–3 mm. long, sharply acute; lower glume 2-toothed at the tip; awn 4·5–7 cm. long, the column pubescent. *Pedicelled spikelets* 4–7 mm. long, shortly mucronate, glabrous except for the ciliate margins; callus 0·2–0·5 mm. long.

MINOR VARIATIONS. In the tiniest plants the pedicelled spikelets are represented by linear scales 3–4 mm. long, and the homogamous spikelets by linear vestiges only 1 mm. long. The racemes thus comprise little more than 1–2 normal sized sessile spikelets.

HABITAT. Upland grassland at 3000 m. on volcanic soils.

DISTRIBUTION. At present known from two highland massifs—Jebel Marra and northern Ethiopia (Map 13).

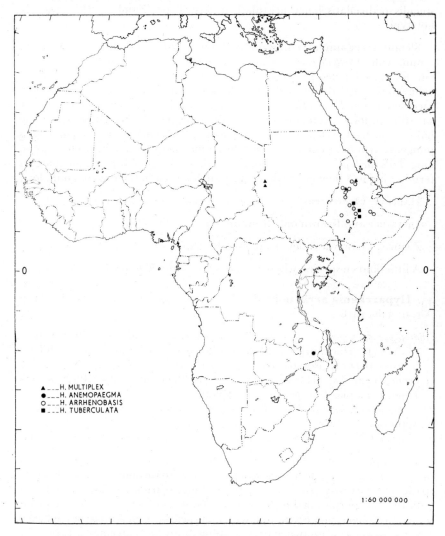

MAP 13. Distribution of *Hyparrhenia multiplex, H. anemopaegma, H. arrhenobasis* and *H. tuberculata*.

SUDAN. Jebel Marra, North of Crater (Sept.), *Blair* 350, 378, *Wickens* 2368, 2470; Jebel Marra, Jebel Ufogo (Dec.), *J. K. Jackson* 3354.

ETHIOPIA. Without locality, *Quartin Dillon & Petit* 68; Bahara district, Sana (Aug.), *Schimper* 1637; Tigre, Mount Arba Tensesa, *Schimper* 349; Gondar (Sept.), *Chiovenda* 1798; Fenter (Aug., Oct.), *Chiovenda* 1485, 2351.

H. multiplex is an annual, often dwarfed, plant, distinguished from the only other annual member of the section by its smaller spikelets and awnless pedicelled spikelets. The characters said to distinguish var. *leiopoda* are neither constant nor reliable, and this variety has therefore been disregarded.

44. **Hyparrhenia anemopaegma** *W. D. Clayton*, sp. nov.; affinis *H. multiplici* (Hochst. ex A. Rich.) Anderss. ex Stapf, sed spiculis aristisque longioribus, spiculis pedicellatis longe aristulatis differt. Type: Zambia, *Astle* 4759 (K, holotype).

Slender erect annual; culms 90 cm. high. *Leaf-blades* up to 15 cm. long and 2 mm. wide. *Spathate panicle* of about 3 raceme-pairs; spatheoles linear, 9–14 cm. long; peduncles ¾ as long as the spatheole, densely bearded with pale yellow hairs near the tip. *Racemes* deflexed, 2 cm. long, 4–6-awned per pair; raceme-bases glabrous. *Homogamous pairs* 1 at the base of each raceme; spikelets 11–16 mm. long, pectinate-ciliate on the margins, shortly pilose with white hairs on the back. *Sessile spikelets* 9 mm. long, pubescent; callus 2·5–3 mm. long, pungent; awn 9–11 cm. long, the column hirtellous with hairs up to 1·5 mm. long. *Pedicelled spikelets* 10–11 mm. long, pilose with shortly pectinate margins, terminating in a slender bristle 11–22 mm. long; callus oblong, 0·5 mm. long.

HABITAT. Open areas in *Erythrophleum* woodland on sandy soil.

DISTRIBUTION. Known only from the type-locality (Map 13, p. 153).

ZAMBIA. Luangwa valley, Kapamba R. (Apr.), *Astle* 4759.

A little-known species, whose relatives are confined to the Ethiopian region.

45. **Hyparrhenia arrhenobasis** *(Hochst. ex Steud.) Stapf* in Prain, Fl. Trop. Afr. 9: 348 (1918).

Andropogon arrhenobasis ('*arrhenobrasis*') Hochst. ex Steud., Syn. Pl. Glum. 1: 385 (1854); Hack. in DC., Monogr. Phan. 6: 626 (1889). Type: Ethiopia, *Schimper* 1821 (K, isotype).
A. papillipes var. *major* Hochst. ex Steud., Syn. Pl. Glum. 1: 385 (1854), *in synon.* Type: Ethiopia, *Schimper* 1054 (P, holotype).
Heteropogon arrhenobasis (Hochst. ex Steud.) Anderss. in Schweinf., Beitr. Fl. Aethiop.: 310 (1867).
Sorghum arrhenobasis (Hochst. ex Steud.) Kuntze, Rev. Gen. Pl. 2: 791 (1891).

Tufted perennial; culms 30–160 cm. high. *Leaf-blades* 10–30 cm. long and up to 5 mm. wide, glaucous. *Spathate panicle* of 2–10 raceme-pairs, rarely more; spatheoles narrowly lanceolate, 4–6 cm. long, scarious, yellowish or tinged with red and purple; peduncles usually about ½ as long as the spatheole but sometimes longer, pilose with yellowish hairs near the tip. *Racemes* not deflexed, 1·5–2·5 cm. long, 7–17-awned per pair, often branched; raceme-bases bearded with fulvous hairs, the upper 1–2 mm. long. *Homogamous pairs* 1 or 2 at the base of both racemes; spikelets 7–12 mm. long, greenish-yellow, pectinate-ciliate on the margins without conspicuous tubercles, otherwise glabrous or sparsely pilose. *Sessile spikelets* 6–7 mm. long, hirsute with light brown hairs or merely hispidulous on the nerves; callus 1 mm. long, acute; lower glume often concave across the back, pallid, leathery; awn 2·5–4 cm. long, the column hispid with hairs 0·3–0·5 mm. long. *Pedicelled spikelets* 7–9 mm. long, acuminate to an awn-point up to 3 mm. long, hirsute or merely hispidulous on the nerves, stiffly ciliate on the margins; callus 0·2 mm. long.

MINOR VARIATIONS. The raceme-bases are typically short and bearded, but occasionally the upper may be almost glabrous, subterete, and up to 2·5 mm. long. The racemes themselves can be hirsute or almost glabrous, but this seems to be of no taxonomic importance.

CHROMOSOME NUMBER. Ethiopia, 2n = 22 (Gould, 1956)!.

HABITAT. Upland grassland at about 3000 m. Edaphic conditions variously described as black clay, wet grassland and dry hillside.

DISTRIBUTION. Ethiopian highlands (Map 13, p. 153).

ETHIOPIA. Adowa, *Schimper* 1054; Tigre, Mettaro, *Schimper* 1010; Schire (Oct.), *Schimper* 1821; Gondar (Oct.), *Chiovenda* 2349; Addis Ababa (Nov.), *Mooney* 5010; Nadda (Oct.), *Mooney* 6102; Gara Mullata (June), *Burger* 2938.

The branched racemes are curious. It is a phenomenon which can occasionally be observed as an abnormality in one or two species, but in *H. arrhenobasis* it is almost a regular feature.

46. **Hyparrhenia tuberculata** *W. D. Clayton*, sp. nov.; affinis *H. arrhenobasi* (Hochst. ex Steud.) Stapf, sed spiculis homogamis tuberculatis, callo longiore, aristis robustioribus pilosioribus differt. Type: Ethiopia, *Mooney* 6284 (K, holotype).

Caespitose perennial; culms 60 cm. high. *Leaf-blades* 10–15 cm. long and 2–3 mm. wide, glaucous. *Spathate panicle* of 2–4 raceme-pairs; spatheoles narrowly lanceolate, 5–9 cm. long, scarious, brown, with or without a short terminal lamina; peduncles about ¾ as long as the spatheole, sometimes longer, pilose with yellow hairs near the tip. *Racemes* not deflexed, 1·5–2 cm. long, 6–8-awned per pair; raceme-bases bearded with yellow hairs, the upper about 2 mm. long. *Homogamous pairs* 2 at the base of each raceme; spikelets 7–12 mm. long, purplish, pectinate-ciliate with tubercle-based bristles encrusting the margins, tuberculate-pilose on the back. *Sessile spikelets* 7–7·5 mm. long, the lower glume beset with short stout prickle-hairs; callus 1·5–2 mm. long, pungent; awn 4–5 cm. long, the column hirtellous with hairs 0·7–1·5 mm. long. *Pedicelled spikelets* 7–8 mm. long, glabrous or scabrid, except for the margins, and sometimes the middle nerve, which are pectinate-ciliate; callus 0·2–0·4 mm. long.

ILLUSTRATION. Fig. 31 (p. 156).

HABITAT. Dry grassland at 2500–3000 m. altitude.

DISTRIBUTION. Confined to the Ethiopian highlands (Map 13, p. 153).

ETHIOPIA. Gojjam, Upper Godeb valley (Aug.), *Lythgoe & Evans* 590; Djem-djem forest (Sept.–Oct.), *Omer-Cooper*; Addis Ababa (Oct.), *Mooney* 8172; Mt. Yerer, 38°55′E., 8°50′N. (Nov.), *Mooney* 6284.

Very closely related to *H. arrhenobasis*; from which it is distinguished by its tuberculate homogamous spikelets, longer callus, and stouter and hairier awns. These characters seem sufficiently distinct at present; it remains to be seen whether they will remain disjunct when more copious material is available.

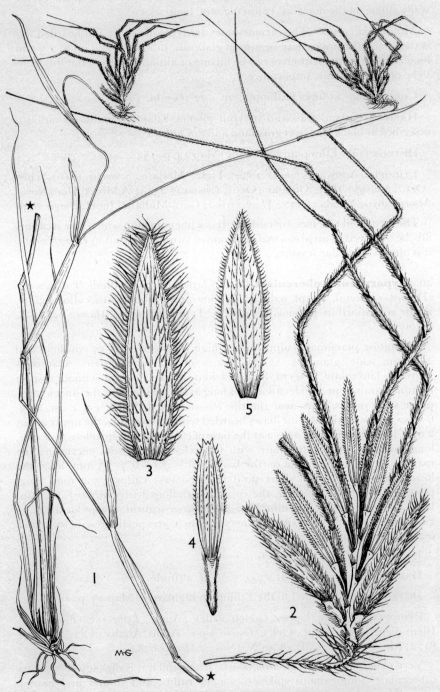

Fig. 31. *Hyparrhenia tuberculata.* **1,** habit, ×⅔; **a,** raceme, ×3. Homogamous spikelet. **3,** lower glume, ×5. Sessile spikelet. **4,** lower glume including callus, ×5. Pedicelled spikelet. **5,** lower glume, ×5. All drawn from *Mooney* 8172.

VI. Sect. **Apogonia** *Stapf* in Prain, Fl. Trop. Afr. 9: 293 (1918); Pilger in Engl., Nat. Pflanzenfam. 14e: 174 (1940). Lectotype species: *H. diplandra.*

Hyparrhenia series *Diplandrae* Stapf, *l. c.*: 293 (1918) Type species: *H. diplandra* (Hack.) Stapf.

Hyparrhenia series *Involucratae* Stapf, *l. c.*: 294 (1918). Type species: *H. involucrata* Stapf.

Hyparrhenia sect. *Dybowskia* (Stapf) Roberty in Boissiera 8: 107 (1960). Type species: *Andropogon seretii* De Wild. (=*Hyparrhenia dybowskii* (Franch.) Roberty).

Tall annuals or perennials, 2–3 m. high. *Leaf-blades* linear; ligule scarious, rounded or truncate. *Spathate* panicle usually ample and rather open; spatheoles lanceolate to linear-lanceolate, green or tinged with red and purple; peduncles up to ½ as long as the spatheole. *Racemes* tardily deflexed, mostly 2–6-awned per pair but sometimes more, laterally exserted; raceme-bases subequal (except *H. gossweileri*), short, flattened, pubescent in the fork but otherwise glabrous (rarely with a few bristles at the tip), the tips truncate and lacking appendages. *Homogamous pairs* 2 at the base of each raceme (except *H. gossweileri*), forming a kind of involucre; spikelets narrowly lanceolate, scabrid on the margins above (or sometimes ciliolate in *H. involucrata*), otherwise glabrous, awnless. *Sessile spikelets* 5·5–8 mm. long (longer in *H. dybowskii*); callus sharply acute to pungent; upper lemma lobes 0·2–1(–1·5) mm. long; awns prominent (except *H. mutica*), the column pubescent to villous with white to rufous hairs. *Pedicelled spikelets* narrowly lanceolate, glabrous, usually shortly awned, without an appreciable callus; pedicel tooth obscure.

A small and homogeneous section characterized by the short flattened raceme-bases lacking bristles, and by the two homogamous pairs at the base of each raceme forming a kind of protective involucre. The latter character is absent from *H. gossweileri*, but this species can scarcely be placed elsewhere than adjacent to *H. diplandra.*

Two homogamous pairs at the base of a raceme occur elsewhere in the genus (*H. filipendula, H. multiplex*), but they are not involucral and they are associated with differences in the raceme-base. Moreover, *H. multiplex* has pectinate-ciliate margins to the homogamous spikelets, a feature not found in sect. *Apogonia.*

The section is dominated by *H. diplandra*, a polymorphic species of wide geographical range. *H. dybowskii* is distinctive, and *H. involucrata* can also be separated without difficulty by virtue of its annual habit. The remaining species are very closely related to the central *H. diplandra* complex with which they intergrade, and their worthiness for separate recognition must be, to some extent, conjectural.

47. **Hyparrhenia dybowskii** *(Franch.) Roberty* in Boissiera 9: 107 (1960).

Andropogon dybowskii Franch. in Bull. Soc. Hist. Nat. Autun 8: 334 (1895). Type: Central African Republic, *Dybowski* 585 (P, holotype).

Cymbopogon princeps Stapf in Mém. Soc. Bot. Fr. 8: 104 (1908). Type: Central African Republic, *Chevalier* 5370 (P, holotype).

Andropogon seretii De Wild., Études Fl. Bas Moyen Congo 3: 152 (1910). Type: Congo Republic, *Seret* 308 (BR, holotype).

Dybowskia seretii (De Wild.) Stapf in Prain, Fl. Trop. Afr. 9: 383 (1918).

D. dybowskii (Franch.) Dandy in Journ. Bot. 69: 54 (1931).

Stout annual; culms up to 2 m. high, perhaps more, bearded at the nodes or not. *Leaf-sheaths* glabrous, except for the ciliate margins; ligule up to 4 mm. long; blades up to 30 cm. or more long and 12 mm. wide, glabrous to sparsely hirsute. *Spathate panicle* about 30 cm. long, of 10–20 raceme-pairs clustered about the axis; spatheoles 6–9 cm. long, green, glabrous or sometimes hirsute at the base; peduncles 0·5–4 cm. long, glabrous or sparsely pilose upwards. *Racemes* 3·5 cm. long, 2-awned per pair; raceme-bases subequal, the upper 1–1·5 mm. long. *Homogamous spikelets* 1·5–2·5 cm. long, scabrid along the margins above. *Sessile spikelets* linear oblong, 1·6–2 cm. long, straw-coloured, glabrous to shortly villous; callus 6 mm. long, pungent; awn 12–19 cm. long, the column shortly and densely pubescent with golden hairs. *Pedicelled spikelets* 1·5–1·9 cm. long, terminating in a short bristle 2–5 mm. long.

ILLUSTRATION. De Wild., Études Fl. Bas Moyen Congo 3: t. 40 (1910).

HABITAT. Imperfectly known and contradictory, the habitat being variously given as 'dalles lateritiques', 'savane' and 'environs de mare'.

DISTRIBUTION. Central African Republic and its borders (Map 14, p. 160).

CHAD. Chari, *Martine.*

CENTRAL AFRICAN REPUBLIC. Yalinga to Wadda (Aug.), *Le Testu* 4135; Guiritoungou (Aug.), *Le Testu* 3196; Yalinga (Sept.), *Le Testu* 3277; Vallée de la Tomi, *Chevalier* 5370; Balao, *Dybowski* 585.

CONGO REPUBLIC. Gemena to Bosobolo (Oct.), *Evrard* 102; Magombo (Oct.), *Gérard* 1893; Tukpwe, *Lecomte* 2; Missa to Gongo, *Seret* 308.

H. dybowskii is rendered conspicuous by the extraordinarily large spikelets and stout awns. It is further differentiated by the short pubescence on the column of the awn, for the other annual species has subplumose awns.

Although acknowledging an affinity with *H. involucrata* Stapf (as *H. notolasia*), Stapf (1918) created a new genus—*Dybowskia*—to accommodate this remarkable plant, basing his decision upon the exuberant size of the racemes and the tiny fertile lemma only 2 mm. long. However the racemes differ from those of *H. notolasia* in little except for their grandiose scale, while the small size of the lemma was due to the immaturity of Stapf's specimens; in mature spikelets the lemma is 6–8 mm. long. I therefore have no hesitation in uniting *Dybowskia* with *Hyparrhenia.*

48. **Hyparrhenia involucrata** *Stapf* in Prain, Fl. Trop. Afr. 9: 377 (1918). Type: Nigeria, *Barter* 957 (K, holotype).

Anthistiria barteri Munro ex Oliver in Trans. Linn. Soc. 29: 176 (1875), *nom. nud. in adnot.*, non *Hyparrhenia barteri* (Hack.) Stapf. Based on *Barter* 957.

Androscoepia barteri Anderss. ex Oliver, *l. c.* (1875), *nom. nud. in adnot.*

Hyparrhenia notolosia Stapf in Prain, Fl. Trop. Afr. 9: 377 (1918). Types: Nigeria, *Dalziel* 299 (K, syntype); Central African Republic, *Macleod* (BM, syntype).

48a. var. **involucrata**

Robust annual; culms 2 m. high, glabrous. *Leaf-sheaths* glabrous or with a few long hairs on the margin; ligule 1–2 mm. long, ciliolate; blades up to 40 cm. long and 8 mm. wide, pale green, flaccid, glabrous, although usually with dense smoky grey hairs about the base and sometimes laxly pilose beneath, very scabrid along the margins. *Spathate panicle* narrow, 20–60 cm. long, lax; spatheoles 4–7 cm. long, scarious, pallid tinged with pink and purple, glabrous and borne upon a glabrous ray; peduncle about ½ as long as the spatheole, glabrous. *Racemes* 1·5–2·3 cm. long, 4-awned per pair; raceme-bases subequal, the upper 1·5 mm. long. *Homogamous spikelets* 7–15 mm. long, scaberulous to ciliolate on the margins above. *Sessile spikelets* narrowly lanceolate, 7–8 mm. long; callus 2–2·5 mm. long, pungent; lower glume glabrous to densely white tomentose on the back; awn 7–11 cm. long, the column hirsute with white or rufous hairs up to 1 mm. long; caryopsis 3·5 mm. long. *Pedicelled spikelets* 7–11 mm. long, terminating in a bristle 8–20 mm. long.

MINOR VARIATIONS. The racemes are rarely 3-awned per pair. The basal node of the spatheole is sometimes shortly bearded, adjacent regions of the spatheole and ray usually remaining glabrous (unlike var. *breviseta*).

HABITAT. In savanna, typically on sandy or gravelly soils, shallow soils over ironstone, or stony hillsides.

DISTRIBUTION. Throughout Nigeria, with a few specimens in Ghana, and to the east of the Cameroun mountains (Map 14, p. 160).

GHANA. Bolgatanga to Tamale (Oct.), *Ankrah* GC 20300; Yendi to Zabzugu (Nov.), *Rose Innes* GC 32472.

N. NIGERIA. Funtua to Yashi (Oct.), *Keay* FHI 21147; Zaria, Samaru (Oct.), *Freeman* S168; Plateau Prov. (Oct.), *Lely* P.849; Nupe, *Barter* 957; Lokoja (Oct.), *Dalziel* 299.

S. NIGERIA. Igbetti (Oct.), *Stanfield* FHI 55499; Ago-Are Forest Reserve (Nov.), *Keay* FHI 37680, FHI 37693; Awba hills Forest Reserve (Nov.), *A. P. D. Jones* FHI 7322; Ngwo (Oct.), *A. P. D. Jones* FHI 6783.

CAMEROUN. Maroua (Oct.), *Vaillant* 2548; Mokolo to Guider (Nov.), *Vaillant* 2569; Garoua to Kapsiki (Oct.), *Koechlin* 7357; Meiganga to Ngaundéré (Oct.), *Koechlin* 7208; Poli (Oct.), *Koechlin* 7291.

CENTRAL AFRICAN REPUBLIC. Mbouras (Nov.), *Macleod*; Bouar (Sept.), *Koechlin* 6242.

48b. var. **breviseta** *W. D. Clayton*, var. nov.; a varietate typica arista spicularum pedicellatarum breviore, pari racemorum plerumque aristis duabus instructa, et basi spatheoli et radio suffultente piloso differt. Type: Ghana, *Rose Innes* GC 30676 (K, holotype).

Base of spatheole and summit of supporting ray pilose with white hairs; racemes 2(–4)-awned per pair; pedicelled spikelets terminating in a bristle 1–5 mm. long. Otherwise similar to var. *involucrata*.

CHROMOSOME NUMBER. Ghana, 2n = 40 *Rose Innes* GC 32373!; 2n = 40 *Rose Innes* s.n. from Savelagu!.

HABITAT. As for var. *involucrata*.

DISTRIBUTION. Best known from Ghana and adjacent territories, but also found in Chad and Central African Republic (Map 14, p. 160).

UPPER VOLTA. Leo to Po (Oct.), *Rose Innes* GC 31475.
IVORY COAST. Séguéla (Oct.), *Aké-Assi* 6567.

GHANA. Wa (Oct.), *Rose Innes* GC 32373; Navrongo (Oct.), *Vigne* FH 4640; Nakpanduri (Sept.), *Rose Innes* GC 32119; Tamale to Bolgatanga, Savelagu (Nov.), *Rose Innes*; Yapei ferry (Oct.), *Rose Innes* GC 30676; Kete Krachi to Yendi (Sept.), *Rose Innes* GC 30448.

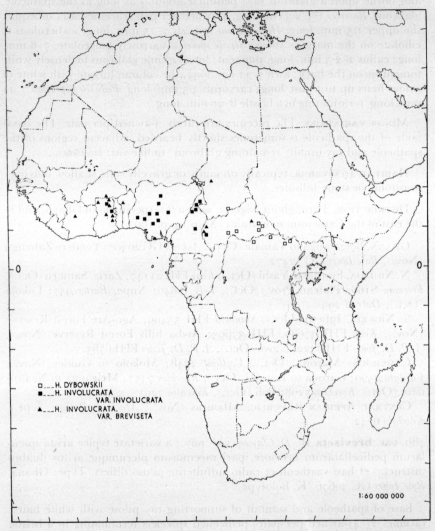

MAP 14. Distribution of *Hyparrhenia dybowskii, H. involucrata* var. *involucrata* and *H. involucrata* var. *breviseta*.

DAHOMEY. Boukombe (Nov.), *Risopoulos* 1259.

CHAD. Moussofoyo (Nov.), *Audru* 1792.

CENTRAL AFRICAN REPUBLIC. Bouar (Sept., Dec.), *Boudet* 1558, *Koechlin* 6243.

In this species the racemes are typically deflexed, but the condition may develop somewhat tardily. Undeflexed specimens in an early stage of development misled Stapf (1918) into assigning *H. involucrata* to a separate series whose racemes were said to be permanently parallel. In fact *H. involucrata* closely resembles *H. subplumosa*, both species having long hairs on the awns. The former differs in its annual habit, long-bristled pedicelled spikelet (or 2-awned raceme-pairs) and generally stouter awn.

Stapf also made a distinction between glabrous and tomentose sessile spikelets, basing his diagnosis of *H. notolasia* upon the latter character. However, this distinction seems to be correlated with no other differences and many intermediates are found. Indeed, a wide range of variation in the indumentum of the sessile spikelet is rather common in this section of *Hyparrhenia*, and I consider it to be of no taxonomic value.

Nevertheless, there are clearly two taxa within the species; differences in number of awns per raceme-pair, spatheole hairiness and length of pedicelled spikelet awn being correlated and geographically segregated. Against this must be set the fact that the correlation of characters is by no means complete, as can be seen from Table 9; and the cumulative effect of small and subtle differences, generally known as the facies, does not make itself apparent. Moreover the geographical evidence is difficult to interpret, and it may be relevant to note that the collections available for examination are heavily biased in favour of north-east Ghana and the Zaria district of Nigeria. My general impression is that we are dealing with the chance local segregation of a few characters, rather than with a complete interruption of gene flow, and I have therefore made the taxonomic distinction at varietal level.

TABLE 9. Frequency distribution of character combinations in *H. involucrata*

	Awns 2 per raceme-pair		Awns 4 per raceme-pair	
	Spatheole base glabrous	Spatheole base hairy	Spatheole base glabrous	Spatheole base hairy
Awn of ped. spikelet 1–5 mm.	0	28	0	5
Awn of ped. spikelet 8–20 mm.	0	2	35	8

Both the description and type of *H. involucrata* are unequivocal, whereas *H. notolasia* is based upon two specimens (one an atypical 3-awned form, the other with the racemes shattered) which have contributed to an unsatisfactory composite description. I have therefore chosen the former name for the species.

49. **Hyparrhenia mutica** *W. D. Clayton*, sp. nov.: *H. diplandrae* (Hack.) Stapf similis, sed spiculis sessilibus muticis differt. Type: Liberia, *Adames* 746 (K, holotype).

Coarse perennial; culms 1·5–2·5 m. high, glabrous. *Leaf-sheaths* glabrous, or pilose near the top, rarely pilose all over; ligule 1 mm. long; blades 30–60 cm. long and up to 10 mm. wide, glabrous or sparsely hirsute, usually with long greyish hairs at the base, scabrid on the margins. *Spathate panicle* narrow, 15–40 cm. long, purplish; spatheoles 2–3·5 cm. long, brownish red, glabrous except for the shortly bearded base; peduncles 0·5–1·5 cm. long, sparsely pubescent above. *Racemes* 1·2–1·5 cm. long, with 4–6 sessile spikelets per pair; raceme-bases subequal, up to 1 mm. long. *Homogamous spikelets* 5–7 mm. long, scabrid along the margins. *Sessile spikelets* narrowly lanceolate to lanceolate-oblong, 5·5–6 mm. long, glabrous or sometimes sparsely pilose; callus up to 1 mm. long, sharply acute; upper lemma lanceolate, 4·5 mm. long, hyaline with ciliate margins, obtuse and mucronulate at the tip; caryopsis 2 mm. long; *Pedicelled spikelets* 5–5·5 mm. long, muticous or with a bristle up to 5 mm. long.

ILLUSTRATION. Fig. 32.

MINOR VARIATIONS. Sometimes a few of the spikelets in a panicle are found to bear very short awns.

CHROMOSOME NUMBER. Sierra Leone, 2n = 60 *Morton* s.n.!

HABITAT. A coarse grass of swampy places.

DISTRIBUTION. Not common, but widely scattered throughout tropical Africa. Its distribution corresponds roughly with that of *H. diplandra*.

SIERRA LEONE. Bintimane (Nov., Dec., Jan.), *J. K. Morton, T. S. Jones* 78, *Jaeger* 492; Loma Mtns. (Nov.), *J. K. Morton* SL 2720.
LIBERIA. Nimba, Beacon Rho (Nov.), *Adames* 746.
GHANA. Esiama (Mar., Dec.), *Rose Innes* GC 30549, GC 30874.
NIGERIA. Gashaka Distr., Gangumi (Dec.), *Latilo & Daramola* FHI 28794.
CAMEROUN. Ippy (Oct.), *Tisserant* 2316; Banda (Mar.), *Raynal* 10586; Ngaundéré to Tibati (Oct.), *Koechlin* 7465.
CENTRAL AFRICAN REPUBLIC. Bouar (Sept.), *Koechlin* 6282.
CONGO REPUBLIC. Parc Nat. Garamba (Nov.), *de Saeger* 1555; Magombo (Oct.), *Gérard* 2019.
SUDAN. Equatoria, Lokota (Sept.), *Myers* 7789.
UGANDA. Masindi (Oct.), *Buechner* 130; Nakasongola (Sept.), *Langdale-Brown* 1321; Serere (Oct.), *Brown* 29; Entebbe, *Fyffe*; L. Nabugabo (July), *Chandler* 1754.
KENYA. Kitale (June), *Bogdan* 3441.
TANZANIA. Nyakato (June), *Gillman* 53; Munene F.R. (Apr.), *Procter & Taylor* 880; Msima (May), *Schlieben* 1008; Songea (Apr.), *Milne-Redhead & Taylor* 9818.
MOZAMBIQUE. Gúruè Mt. (Nov.), *Mendonça* 2129.
MALAWI. Chikwewo (May), *G. Jackson* 828.
ZAMBIA. Abercorn, Nakatali (June), *Bullock* 2955; Mufulira (May), *Eyles* 8392.
ANGOLA. Biula (Apr.), *Gossweiler* 11291; Nova Lisboa, *Gossweiler* 11890.

H. mutica is an unusual awnless species of *Hyparrhenia*, very closely related to *H. diplandra*, of which it might be regarded simply as an awnless form. The scanty cytological evidence currently available indicates that it differs in chromosome number from *H. diplandra*, thereby affording some support for its separation as a new species.

H. mutica is easily confused with specimens of *H. diplandra* in which the florets have been replaced by a fungal infructescence. It is therefore necessary to dissect a spikelet before confirming the identity of this species.

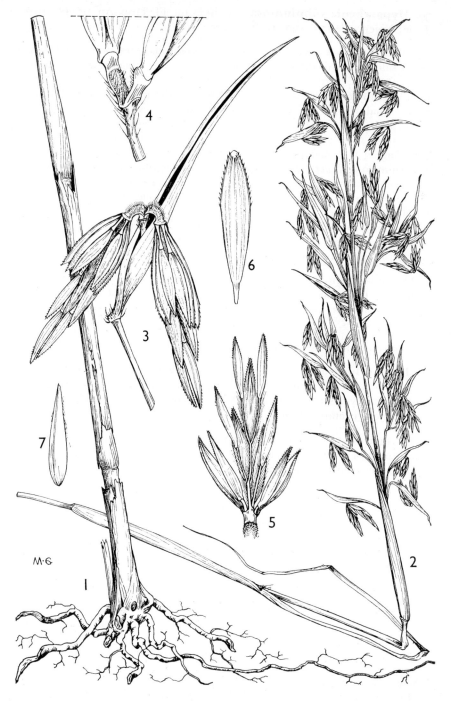

Fig. 32. *Hyparrhenia mutica.* **1,** base of plant showing roots and upright habit of stem, ×⅔; **2,** inflorescence, ×⅔; **3,** raceme-pair, ×4; **4,** raceme-bases, ×12; **5,** single raceme, spread open, ×4. Sessile spikelet. **6,** lower glume including callus, ×6; **7,** upper lemma, ×6. All drawn from *P. Adames* 746.

50. **Hyparrhenia subplumosa** *Stapf* in Prain, Fl. Trop. Afr. 9: 368 (1918).
Type: Nigeria, *Dalziel* 190 (K, lectotype).

Robust perennial; culms 2–3 m. high, glabrous. *Leaf-sheaths* glabrous or
rarely pilose towards the top; ligule up to 3 mm. long; blades 20–60 cm. long
and 3–10 mm. wide, glaucous, glabrous or rarely sparsely hirsute below,
usually with smoky grey hairs near the base, the margins scabrid. *Spathate
panicle* usually large, 20–50 cm. long, loose; spatheoles 3–7 cm. long, glaucous
to purplish, glabrous, bearded or not at the base; peduncles 1–3·5 cm. long,
glabrous or shortly and softly hirsute above, often recurved. *Racemes* 1·5–2·5
cm. long, 3–6-awned per pair; raceme-bases slightly unequal, the upper
1·5–2 mm. long. *Homogamous spikelets* 8–10 mm. long, glabrous or rarely sparsely
pubescent, scabrid on the margins, initially green but becoming purplish.
Sessile spikelets narrowly lanceolate, 6·5–7·5 mm. long, glabrous or less commonly pubescent to tomentose; callus 1·5 mm. long, sharply acute; awn 4·5–
7·5 cm. long, the column subplumose with white or fulvous hairs 0·5–1·3(–1·7)
mm. long; caryopsis 3·5 mm. long. *Pedicelled spikelets* 7–8 mm. long, terminating
in a bristle 2–7 mm. long.

MINOR VARIATIONS. Exceptionally, the upper raceme-base may be up to 3
mm. long. I have seen one specimen in which the upper raceme bears only 1
homogamous pair, and another in which the homogamous pairs are deficient
in both racemes.

CHROMOSOME NUMBER. Ghana, 2n = 40 *Rose Innes* GC 32470!.

HABITAT. In savanna woodland and tree savanna, usually on poor dry soils
such as gravelly hill slopes or sandy soils with an ironstone horizon near the
surface; rarely descending to moister streamside soils. A typical constituent of
the Guinea vegetation type.

DISTRIBUTION. Widespread in west tropical Africa, where it is a common
constituent of savanna. There are also isolated specimens, which apparently
must be referred to this species, scattered through the circum-Congo region
(Map 15).

SENEGAL. Vélingara (Oct.), *Adam* 18619; Tambacounda (Nov.), *Adam*
20031; Tambacounda to Niokolo Koba (Oct.), *Adam* 15638; Niokolo Koba
(Nov.), *Adam* 17070.
    MALI. Sikasso (Sept.), *Adam* 15460.
    UPPER VOLTA. Samandéni (Oct.), *Kmoch* 128; Bobo Dioulasso (Oct.),
Kmoch 147; Ouassa (Oct.), *Rose Innes* GC 31509; Leo (Oct.), *Rose Innes* GC
31074; Leo to Po (Oct.), *Rose Innes* GC 31480.
    GUINEA. Pita (Oct.), *Adam* 12555; Dalaba, *Chevalier* 20185; Timbo (Oct.),
Pobéguin 1799; Baffing valley (Nov.), *Pobéguin* 1812; Siguiri (Oct.), *Jacques-
Félix* 1331.
    SIERRA LEONE. Kambia (Nov.), *Jordan* 689; Mabala (Oct.), *Glanville* 62;
Samaia (May), *N. W. Thomas* 227; Musaia (Dec.), *Deighton* 4444; Mamodia
(Nov.), *Glanville* 329.
    LIBERIA. Nimba (Oct., Nov.), *Adames* 698, 748.
    IVORY COAST. Ferkessedougou (Nov.), *Leeuwenberg* 1999; Séguéla to Vavoua
(Oct.), *Adjanohoun* 961a; Zuénoula to Vavoua (Oct.), *Adjanohoun* 313a;
Toumodi (Apr.), *Leeuwenberg* 3321; Dabou (Nov.), *Leeuwenberg* 1967.

GHANA. Wa to Lawra (Oct.), *Rose Innes* GC 32354; Gambaga (Oct.), *Ankrah* GC 20285; Yendi to Zanzugu (Nov.), *Rose Innes* GC 32470; Elmina (Nov.), *Rose Innes* GC 31171; Achimota (June), *Irvine* 3056; Kpong to Somanya (Nov.), *Rose Innes* GC 31647.

TOGO. Misahohe, *Baumann* 328.

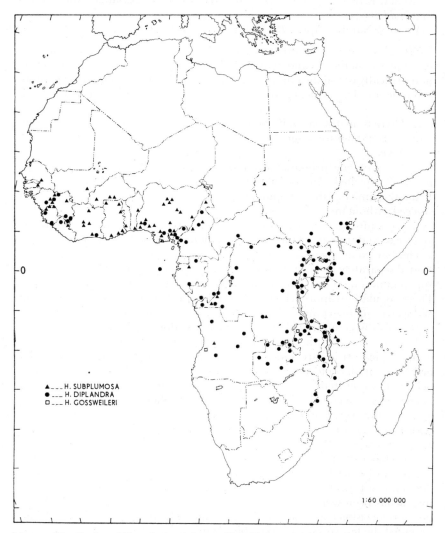

MAP 15. Distribution of *Hyparrhenia subplumosa, H. diplandra* and *H. gossweileri.*

DAHOMEY. Parakou, *Risopoulos* 1294; Cotonou to Allada (Oct.), *Risopoulos* 1234.

NIGERIA. Zaria (Nov.), *Thatcher* S536; Anara F.R. (Oct.), *Keay* FHI 20109; Abinsi, *Dalziel* 901; Olokemeji (Nov.), *Clayton* 566; Akpaka F.R. (Nov.), *Onochie* FHI 40429; Obudu Plateau (Feb.), *Tuley* 489.

CAMEROUN. Mokolo (Nov.), *Vaillant* 2567, 2570; Tibati (Sept.), *Letouzey* 5771; Kapsiki (Oct.), *Koechlin* 7366; Ngaundéré (Oct.), *Koechlin* 7235.

GABON. Booué (Aug.), *J. N. Davies* 257.

CONGO REPUBLIC. Kisantu, *Callens* 178; Leopoldville (Sept.), *Compère* 308; Taka, *Vanderyst* 3410; Gandajika (Apr.), *Risopoulos* 268.

SUDAN. Jebel Marra (Dec.), *J. K. Jackson* 3381.

TANZANIA. Songea (May), *Milne-Redhead & Taylor* 9895.

ZAMBIA. Kawimbe (Feb.), *McCallum Webster* A168; Kasama (Apr.), *Vesey-FitzGerald* 1664.

ANGOLA. Namba (Apr.), *Barbosa & Correia* 9148.

Hyparrhenia subplumosa differs from *H. diplandra* in its looser panicle with longer awns, spatheoles and peduncles, but attempts to analyse the differences more critically reveal the tantalizing difficulties which confront the taxonomist in this genus. The evidence for separating these taxa may be summed up as follows:

1. There is no discontinuity between the variable characters, but the shape of the scatter diagram suggests that two partially superposed populations may be involved.

2. The different chromosome numbers confirm that at least two populations are to be expected. An adequate number of counts would no doubt help to define the limits of these populations, but for the present we must perforce proceed without them.

3. Field observations lead to a firm subjective impression that the plants to the east and the west of the Cameroun mountain range are different.

The last point suggests a fresh approach, and when scatter diagrams are plotted separately for west Africa, and for east and central Africa, it is at once apparent that there are two distinct populations (Figs. 33 & 34, p. 167 & 168). These populations approach very closely, but the length of hairs on the column of the awn (up to 0·4 mm. for *H. diplandra*, 0·5 mm. or more for *H. subplumosa*) provides a distinguishing character which is seldom equivocal.

The full extent of the dilemma becomes apparent when specimens from south tropical Africa are added (Fig. 35, p. 169). Their variation in awn and peduncle length conforms to that of *H. diplandra*, to which species they clearly belong, but their tendency to have longer awn hairs means that they can no longer be satisfactorily distinguished from peripheral specimens of *H. subplumosa*. However, I do not believe that the presence of this form in southern Africa is a good reason for uniting the distinctive West African population with *H. diplandra*.

To what extent do these species transgress into each others' territory? There can be no doubt that *H. diplandra* occurs to a limited extent in West Africa. The trans-Cameroun distribution of *H. subplumosa* is more nebulous for, as has been seen, the species are difficult to separate in southern Africa, and any decision must necessarily be somewhat arbitrary. In short I have accepted the few specimens resembling typical *H. subplumosa*, and consigned the rest to *H. diplandra*.

Stapf's (1918) original description was supported by a large number of syntypes, and I have felt it advisable to nominate a lectotype from among them.

51. **Hyparrhenia diplandra** *(Hack.) Stapf* in Prain, Fl. Trop. Afr. 9: 368 (1918).

Andropogon diplandrus Hack. in Flora 68: 123 (1885); & in DC., Monogr. Phan. 6: 627 (1889). Types: Sudan, *Schweinfurth* 2002, 2094 (both K, isosyntypes).

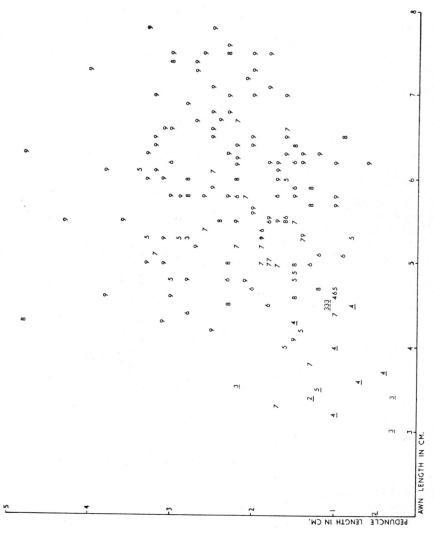

PEDUNCLE LENGTH IN CM.

AWN LENGTH IN CM.

Fig. 33. Lengths of awn and peduncle for specimens of *Hyparrhenia subplumosa* from West Africa. Numbers represent length of awn hairs in units of 0·1 mm. (9 = 0·9 mm. or more). Underline indicates specimens accepted as *H. diplandra*.

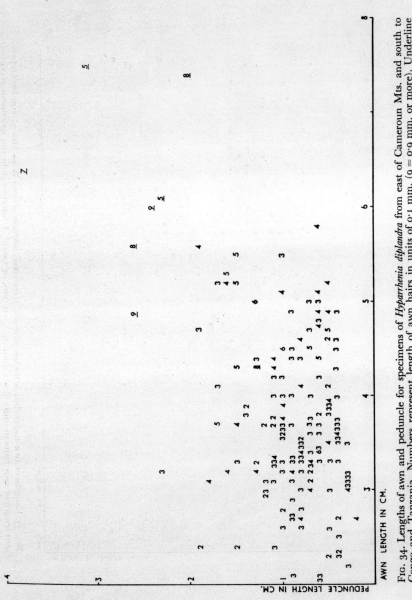

Fig. 34. Lengths of awn and peduncle for specimens of *Hyparrhenia diplandra* from east of Cameroun Mts. and south to Congo and Tanzania. Numbers represent length of awn hairs in units of 0·1 mm. (9 = 0·9 mm. or more). Underline indicates specimens accepted as *H. subplumosa*.

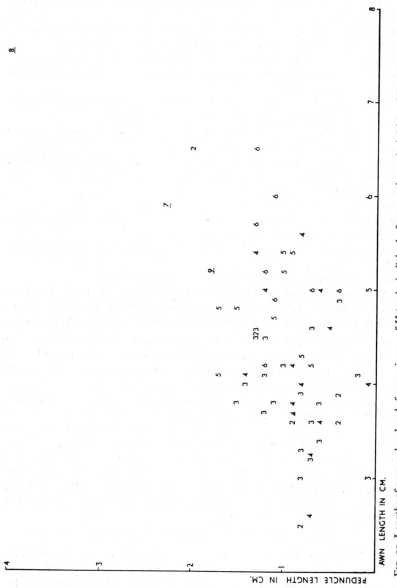

FIG. 35. Lengths of awn and peduncle for specimens of *Hyparrhenia diplandra* from south tropical Africa. Numbers represent length of awn hairs in units of 0·1 mm. (9 = 0·9 mm. or more). Underline indicates specimens accepted as *H. subplumosa*.

Sorghum diplandrum (Hack.) Kuntze, Rev. Gen. Pl. 2: 791 (1891).

Andropogon osikensis Franch. in Bull. Soc. Hist. Nat. Autun 8: 332 (1895). Type: Republic of Congo, *Brazza & Thollon* 233 (P, holotype).

A. pachyneuros Franch., *l. c.*: 333 (1895). Type: Republic of Congo, *Thollon* 777 (P, holotype).

A. obscurus K. Schum. in Engl., Bot. Jahrb. 24: 330 (1897). Type: Cameroun, *Zenker* 544 (K, isotype).

Cymbopogon phoenix Rendle, Cat. Afr. Pl. Welw. 2: 156 (1899). Types: Angola, *Welwitsch* 7193 (BM, syntype), 7226 (BM, syntype; K, isosyntype).

Andropogon phoenix (Rendle) K. Schum. in Just, Bot. Jahresb. 27 (1): 454 (1901).

A. vulgaris Vanderyst in Bull. Agric. Congo Belge 9: 243 (1918), *nom. prov.*; & *l. c.* 11: 145 (1920), *in synon.*

A. vulgaris var. *glaucus* Vanderyst in Bull. Agric. Congo Belge 9: 243 (1918), *nom. prov.*; & in Bull. Soc. Bot. Belge 55: 45 (1923), *in synon.*

Hyparrhenia pachystachya Stapf in Prain, Fl. Trop. Afr. 9: 370 (1918). Types: Zambia, *Macaulay* (K, syntype); Rhodesia, *Swynnerton* 993 (K, syntype).

Cymbopogon diplandrus (Hack.) De Wild. in Bull. Jard. Bot. Brux. 6: 11 (1919).

C. kapandensis De Wild., *l. c.*: 13 (1919). Type: Congo Republic, *Homblé* 991 (BR, holotype; K, isotype).

Andropogon kapandensis De Wild., *l. c.* (1919), *in synon.*

Cymbopogon eberhardtii A. Camus in Bull. Mus. Hist. Nat., Paris 25: 133 (1919). Type: Vietnam, *Eberhardt* 1857 (P, holotype).

Hyparrhenia diplandra var. *major* Vanderyst in Bull. Agric. Congo Belge 11: 145 (1920), *nom. prov.*

Andropogon vulgaris var. *major* Vanderyst, *l. c.* (1920), *in synon.*

Hyparrhenia takaensis Vanderyst in Bull. Soc. Bot. Belge 55: 45 (1923), *nom. prov.*

Andropogon eberhardtii (A. Camus) Merrill in Lingnan Sci. Journ. 5: 27 (1928).

Hyparrhenia eberhardtii (A. Camus) Hitchc. in Lingnan Sci. Journ. 7: 247 (1931).

Coarse perennial; culms 2–3 m. high, glabrous. *Leaf-sheaths* glabrous, or pilose towards the top, rarely pilose throughout; ligule 2 mm. long; blades 20–60 cm. long and 3–10 mm. wide, glabrous or sometimes sparsely hirsute beneath, usually with long grey hairs at the base. *Spathate panicle* 20–40 cm. long, usually purplish; spatheoles 2–4·5 cm. long, brownish red, glabrous, bearded or not at the base; peduncles 0·3–1·5 cm. long, glabrous or shortly hirsute above. *Racemes* 1·5–2(–2·5) cm. long, (3–)4–6(–9)-awned per pair; raceme-bases subequal, the upper 1–2 mm. long. *Homogamous spikelets* 7–9 mm. long, scabrid on the margins. *Sessile spikelets* narrowly lanceolate to lanceolate oblong, 6–8 mm. long, rarely shorter, glabrous to pubescent; callus 1·5 mm. long, sharply acute; awn 2–5·5 cm. long, the column with white or fulvous hairs 0·2–0·4(–0·5) mm. long. *Pedicelled spikelets* 5–7·5 mm. long, with or without a terminal bristle up to 5 mm. long.

ILLUSTRATION. Robyns, Fl. Agrost. Congo Belge 1: 195, t. 14 (1929).

MINOR VARIATIONS. Panicles in which some of the racemes are supported by only 1 homogamous pair are very occasionally found; even rarer are specimens (such as *Troupin* 826 from Congo) in which all the racemes have only one homogamous pair. In other respects these specimens resemble typical *H. diplandra*, and there is no good reason for separating them.

Occasionally the lowest internodes of the raceme are very short and the spikelets therefore tightly bunched. Very rarely the upper raceme-base bears two unequal, and more or less deformed, racemes.

The spikelets are frequently infected by the smut fungi *Sphacelotheca barcinonensis* Riofrio and *S. congoensis* (H. & P. Sydow) Wakefield.

CHROMOSOME NUMBER. Sierra Leone, 2n = 20 *J. K. Morton* 447!.

HABITAT. The species occupies a wide range of habitats from swamps and drainage hollows to stony hillsides and ironstone outcrops.

DISTRIBUTION. Tropical Africa to the south and east of the Cameroun Mts., with a noticeably circum-Congo distribution; intrudes across the Cameroun Mountains into the eastern border of Nigeria, and also into the Guinea highlands (Map 15, p. 165). It is also found in Thailand, Vietnam and Indonesia.

CHINA. Hainan, Nodoa (Nov.), *McClure* 7915.

THAILAND. Poo Kradeng (Mar., Nov.), *Larsen* 2280, *Smitinand* 2107.

VIETNAM. Lang Bian (Feb.), *Chevalier* 30664.

CELEBES. Sae (July), *Eyma* 1202; Singkalong (Aug.), *Eyma* 1468.

GUINEA. Dalaba (Oct.), *Adam* 12673; Baffing (Oct.), *Pobéguin* 1803; Konkouré (Mar.), *Pitot* 279; Macenta (Oct.), *Baldwin* 9750; Nimba (Oct.), *Schnell* 3865.

SIERRA LEONE. Kambia (Dec.), *Deighton* 877; Senehun (Apr.), *Morton & Gledhill* SL 1964; Loma Mansa (Dec.), *J. K. Morton* SL 447.

LIBERIA. Mt. Loma (Nov.), *Jaeger* 499; Gbarnga (Dec.), *Baldwin* 10538; Lamco base camp (Apr.), *Harley* 2211; Nimba (Dec., Jan.), *Adam* 20207, 20637.

IVORY COAST. Bingerville (Dec.), *Chevalier* 20091, *Hédin*.

NIGERIA. Oyo (Oct.), *Stanfield*; Mambila plateau, Mandaga (Jan.), *Latilo & Daramola* FHI 28993; Enugu (May), *A. P. D. Jones* FHI 18862; Ogoja Prov., *Roseveare* 16/30a; Obudu (Apr.), *Wheeler-Haines* 370.

CAMEROUN. Nkambe (Feb.), *Charter* FHI 36981; Wum (Nov.), *Brunt* 875; Bafut-Ngemba F.R. (Mar.), *Hepper* 2242; Bamenda (Dec.), *Baldwin* 13838; Ndu (Jan.), *Keay* FHI 28432; Nkongsamba (Dec.), *Baldwin* 13874; Yaoundé, *Zenker* 544; Nango (Oct.), *Vaillant* 2148; Mikila (Oct.), *Letouzey* 6135; Meianga to Ngaundéré (Oct.), *Koechlin* 7176.

CENTRAL AFRICAN REPUBLIC. Krébedjé (Oct.), *Chevalier* 5695; N'dimbi (Oct.), *Audru & Boudet* 3267; Waka (Oct.), *Tisserant* 2698; Mbaiki (Nov.), *Tessmann* 2099.

GABON. St. Martin (May), *Walker*; Ndendé (May), *Guillemet*.

REPUBLIC OF CONGO. Ogooné, *Thollon* 777; Osika (June), *Brazza & Thollon* 233; Brazzaville (July), *Chevalier* 27363; Loudima (May), *Bonnin* 406; Madigou (May), *Koechlin* 2667.

SÃO THOMÉ. Without locality, *Chevalier* 13785.

CONGO REPUBLIC. Uele (Nov.), *de Saeger* 1518; Coquillatville (Sept.), *Robyns* 790; Plaine de la Semliki (Aug.), *Louis* 5526; Kivu, Kalehe (Feb.), *Léonard* 3031; Kisantu (May), *Callens* 3588; Stanleypool (Apr.), *Hens* 25, *Vanderyst* 3838; Kapanda valley (Dec.), *Homblé* 991; Elisabethville (Apr.), *Gathy* 753.

RWANDA. Bukavu to Astrida (May), *Troupin* 9916; Astrida (Apr.), *Van der Ben* 1536.

BURUNDI. Karuzi (Mar., Apr.), *Van der Ben* 1969, 2032.

SUDAN. Seriba Ghattas (Nov.), *Schweinfurth* 2602; Tondy R. (Nov.), *Schweinfurth* 2644; Lado (Aug.), *Cartwright* 19; Yei, Mt. Gumbiri (June), *J. K. Jackson* 4271; Imatong Mts. (June), *Andrews* A1861.

ETHIOPIA. Jimma (Oct.), *R. B. Stewart* 183; Gosab valley, Tadesse farm (Dec.), *Siegenthaler* 1582; Gojeb valley (Nov.), *Mooney* 8621; Agheremariam (Nov.), *Gillett* 14488.

UGANDA. West Nile, Mt. Otze (June), *A. S. Thomas* 1966; Masindi, *Lewys-Lloyd* 13; Serere (Oct.), *Brown* 28; Karamoja, Tumu forest (Nov.), *A. S. Thomas* 3233; Kampala (Mar.), *Michelmore* 1279; Mbarara to Kabale (Mar.), *Michelmore* 1347.

KENYA. Bungoma (June), *Bogdan* 4061; Kitale (June), *Bogdan* 3442; Nyeri (Nov.), *Lyne-Watt* 1123; Embu (Dec.), *Bogdan* 2688; Kisii (May), *Tateoka* 3070.

TANZANIA. Bukoba (Feb.), *Haarer* 2508; Musoma (June), *Emson* 272; Ufipa, Sumbawanga (Apr.), *Vesey-FitzGerald* 1712; Mbeya, Mbozi, *Jacobsen* 7; Njombe, *Emson* 125.

MOZAMBIQUE. Gúruè (Apr.), *Torre* 5174; Massingire (May), *Torre* 5377; Angónia (May), *Mendonça* 4186; Dondo to Beira (May), *Mendonça* 4430; Spungabera (June), *Torre* 4263.

MALAWI. Misuku (June), *G. Jackson* 566; Karonga (July), *Whyte*; Fort Manning (Jan.), *G. Jackson* 720; Lilongwe (Apr.), *G. Jackson* 2222.

ZAMBIA. Abercorn (Apr.), *Trapnell* 1775; Luwingu (May), *Astle* 621; Mwinilunga (Oct.), *Milne-Redhead* 2604; Mufulira (May), *Eyles* 8402; Broken Hill (Apr.), *Hinds* 106; Mumbwa, *Macaulay*.

RHODESIA. Upper Buzi R. (Apr.), *Swynnerton* 993; Melsetter (Mar.), *Brain* 10586.

ANGOLA. Uije (Jan.), *Gossweiler* 7370; Kela, *Gossweiler* 9559; Muriege (Apr.), *Gossweiler* 11711; Nova Lisboa (May), *Gossweiler* 10767; Vila Luzo (May), *Gossweiler* 11258.

A widespread and rather variable species, as the long synonymy will testify. Moreover it embraces a wide range of habitats, far greater than *H. subplumosa*. Nevertheless there seem to be no discontinuities within the variable characters (Fig. 34, p. 168), nor correlations between them, and I am driven to conclude that only a single species should be recognized. It is quite possible that a variable species of this sort should embrace several different chromosome numbers, and the interim assumption that it forms the base of a polyploid series extending into *H. subplumosa* and *H. mutica* may prove to be an oversimplification when further counts are available.

Some specimens from the Sudan look very like *H. dregeana*, a species with bearded raceme-bases, 1 homogamous pair on the lower raceme only, and shorter and more numerous awns. There is thus little chance of confusion, despite the superficial similarities, even with specimens of *H. diplandra* deficient in homogamous spikelets.

52. **Hyparrhenia gossweileri** *Stapf* in Prain, Fl. Trop. Afr. 9: 371 (1918). Type: Angola, *Gossweiler* 3085 (K, holotype).

Cymbopogon bequaertii De Wild. in Bull. Jard. Bot. Brux. 6: 8 (1919). Type: Congo Republic, *Bequaert* 306 (K, isotype).
Andropogon bequaertii De Wild., *l. c.*, *in synon.*
Hyparrhenia bequaertii (De Wild.) Robyns, Fl. Agrost. Congo Belge 1: 197 (1929).

Erect perennial; culms 2 m. high, glabrous. *Leaf-sheaths* glabrous or shortly pubescent towards the top and with smoky grey hairs at the mouth; ligule 1 mm. long, ciliolate; blades up to 30 cm. long and 3–6 mm. wide, rigid, glabrous, the margins scabrid. *Spathate panicle* scanty, 30–60 cm. long, lax; spatheoles 4–7 cm. long, reddish purple, glabrous, bearded or not at the base; peduncles 1·5–3 cm. long, glabrous or shortly hirsute above. *Racemes* 2·5–3 cm. long, 6–12-awned per pair; raceme-bases unequal, the lower 1 mm. long, the upper 3–4 mm. long. *Homogamous pairs* 1 at the base of each raceme; spikelets 8–10 mm. long, scabrid on the margins near the tip. *Sessile spikelets* narrowly lanceolate, 8 mm. long, glabrous; callus 2 mm. long, sharply acute; awn 3·5–5 cm. long, the column pubescent with fulvous hairs 0·2–0·3 mm. long. *Pedicelled spikelets* 7–8 mm. long, muticous.

HABITAT. In savanna woodland; uncommon.

DISTRIBUTION. A species of limited distribution, centred on Katanga (Map 15, p. 165); a single outlying specimen from Ethiopia.

CONGO REPUBLIC. Elizabethville (Apr.), *Bequaert* 306.
ETHIOPIA. Jimma (Oct.), *Mooney* 5887.
TANZANIA. Mbeya (Apr.), *Boaler* 537.
ZAMBIA. Luwingu (May), *Astle* 625; Chisinga ranch (May), *Astle* 3023.
ANGOLA. Munongue (Apr.), *Gossweiler* 3085.

H. gossweileri closely resembles *H. diplandra*, but has only 1 homogamous pair at the base of each raceme. This character is not absolutely constant in *H. diplandra*, but further distinguishing characters are provided by the unusually long upper raceme-base, and by the longer racemes with more awns per pair. Although obviously a peripheral fragment, these characters seem sufficient to justify the recognition of *H. gossweileri* as a separate species.

DOUBTFUL AND EXCLUDED NAMES

Hyparrhenia abyssinica (Hochst. ex A. Rich.) Rob. in Boissiera 9: 108 (1960) = **Exotheca abyssinica** (*Hochst. ex A. Rich.*) *Anderss.*

H. archaelmyandra Jac.-Fél. in Journ. Agric. Trop. 1: 48 (1954) = **Elymandra archaelymandra** (*Jac.-Fél.*) *W. D. Clayton.*

H. baddadae Chiov., Fl. Somala 2: 440 (1932). A teratological specimen of **Elymandra,** probably *E. grallata* (Stapf) W. D. Clayton.

H. cornucopiae (Hack.) Stapf in Prain, Fl. Trop. Afr. 9: 378 (1918) = **Hyperthelia cornucopiae** (*Hack.*) *W. D. Clayton.*

H. dissoluta (Nees ex Steud.) C. E. Hubbard in Hutch. & Dalz., Fl. W. Trop. Afr. 2: 591 (1936) = **Hyperthelia dissoluta** (*Nees ex Steud.*) *W. D. Clayton.*

H. djalonica Jac.-Fél. in Journ. Agric. Trop. 1: 51 (1954) = **Parahyparrhenia annua** (*Hack.*) *W. D. Clayton.*

H. edulis C. E. Hubbard in Hook., Ic. Pl. 35, t 3495 (1950) = **Hyperthelia edulis** (*C. E. Hubbard*) *W. D. Clayton.*

H. eylesii C. E. Hubbard in Bull. Misc. Inf. Kew 1928: 37 (1928) = **Elymandra grallata** (*Stapf*) *W. D. Clayton.*

H. genniamia Anderss. in Schweinf., Beitr. Fl. Aeth.: 300 (1867), *nom. nud. & comb. invalid.* There is no clue to the identity of this name.

H. grallata Stapf in Prain, Fl. Trop. Afr. 9: 320 (1918) = **Elymandra grallata** (*Stapf*) *W. D. Clayton.*

H. jaegerana (A. Camus) Rob. in Boissiera 9: 109 (1960) = **Parahyparrhenia annua** (*Hack.*) *W. D. Clayton.*

12

H. lithophila (Trin.) Pilger in Engl. & Prantl, Nat. Pflanzenfam. 14e: 174 (1940) = **Elymandra lithophila** (*Trin.*) *W. D. Clayton*.

H. macrolepis (Hack.) Stapf in Prain, Fl. Trop. Afr. 9: 328 (1918) = **Hyperthelia macrolepis** (*Hack.*) *W. D. Clayton*.

H. monathera (A. Rich.) Schweinf., Beitr. Fl. Aeth.: 300, 310 (1867), *comb. invalid.* = **Exotheca abyssinica** (*Hochst. ex A. Rich.*) *Anderss.*

H. pendula Peter in Fedde, Rep. Sp. Nov., Beih. 40 (1): 374 (1936). Based on *Andropogon pendulus* Peter *l. c.*, Anh.: 16, t. 16 (1929), *non* Nees ex Steud. (1854). Type: Tanzania, *Peter* 37879 (B, destroyed during second world war). The figure, and the dimensions given in the description, do not agree with any known species of *Hyparrhenia*. Peter himself placed the plant between *H. rufa* and *H. nyassae*, and it is possible that he was describing an abnormal specimen of one of these. At any rate the species cannot now be identified.

H. pusilla (Hook. f.) Stapf in Prain., Fl. Trop. Afr. 9: 379 (1918) = **Andropogon pusillus** *Hook. f.*

H. ruprechtii Fourn., Mex. Pl. 2: 67 (1886) = **Hyperthelia dissoluta** (*Nees ex Steud.*) *W. D. Clayton*.

H. sulcata Jac.-Fél. in Journ. Agric. Trop. 1: 54 (1954) = **Parahyparrhenia annua** (*Hack.*) *W. D. Clayton*.

References

Adjanohoun, E. & Clayton, W. D. (1963). Un *Andropogon* nouveau de la section *Piestium* (Graminées). Adansonia 3: 401–403.

Agreda, O. & Cuany, R. L. (1962). Efetos fotoperiodica y fecha de floracion en Jaragua. Turrialba 12: 146–149.

Anderson, E. (1948). Hybridization of the habitat. Evolution 2: 1–9.

—— (1949). Introgressive hybridization. New York.

Andersson, N. J. (1856). Monographiae Andropogonearum, I. Anthistirieae. Nov. Act. Soc. Sc. Upsal. ser. 3, 2: 229–255.

Bentham, G. & Hooker, J. D. (1883). Genera plantarum 3: 1133–1135. London.

Bonsma, J. C. (1949). Breeding cattle for increased adaptability to tropical and subtropical environments. Journ. Agric. Res. 29: 204–221.

Boughey, A. S. *et al.* (1964). Antibiotic reactions between African savanna species. Nature 203: 1302–1303.

Brown, W. V. & Emery, W. H. P. (1957a). Persistent nucleoli in grass systematics. Amer. Journ. Bot. 44: 585–590.

—— & —— (1957b). Some South African apomictic grasses. Journ. S. Afr. Bot. 23: 123–125.

—— & —— (1958). Apomixis in the *Gramineae*: *Panicoideae*. Amer. Journ. Bot. 45: 253–263.

Celarier, R. P. (1956). Additional evidence for 5 as the basic chromosome number of the *Andropogoneae*. Rhodora 58: 135–143.

—— & Harlan, J. R. (1957). Annual report of progress in forage crops research in Oklahoma, 1957. Mimeographed, Stillwater.

Clayton, W. D. (1957). The swamps and sand dunes of Hadejia. Niger. Geogr. Journ. 1: 31–37.

—— (1958a). Secondary vegetation and the transition to savanna near Ibadan, Nigeria. Journ. Ecol. 46: 217–238.

Clayton, W. D. (1958b). Erosion surfaces in Kabba Province, Nigeria. Journ. W. Afr. Sci. Ass. 4: 141–149.

—— (1959). Report on the vegetation of map sheet 181 (N. W. Mokwa). Soil Survey Section, Regional Research Station, Northern Nigeria, Bull. No. 4: 12–17.

—— (1961). Derived savanna in Kabba Province, Nigeria. Journ. Ecol. 49: 595–604.

—— (1963). The vegetation of Katsina Province, Nigeria. Journ. Ecol. 51: 345–351.

—— (1964). Studies in the *Gramineae*: V. Kew Bull. 17: 465–470.

—— (1965). Studies in the *Gramineae*: VII. Kew Bull. 19: 451–456.

—— (1966a). Studies in the *Gramineae*: VIII. Kew Bull. 20: 73–76.

—— (1966b). Vegetation ripples near Gummi, Nigeria. Journ. Ecol. 54: 415–417.

—— (1966c). Studies in the *Gramineae*: X. Kew Bull. 20: 275–285.

—— (1966d). Studies in the *Gramineae*: XI. Kew Bull. 20: 287–293.

—— (1966e). Studies in the *Gramineae*: XII. Kew Bull. 20: 433–449.

—— (1967). *Andropogon pteropholis*. Hook., Ic. Pl., t. 3644.

Compère, P. (1963). Les noms provisoires du prodrome d'agrostologie de Vanderyst. Bull. Jard. Bot. Brux. 33: 383–398.

Corbett, D. C. M. (1966). Central African nematodes. III. *Anguina hyparrheniae* n. sp. associated with 'witches' broom' of *Hyparrhenia sp.* Nematologica 12: 280–286.

Culwick, G. M. (1950). A dietary survey among the Zande of the south west Sudan. Khartoum.

Freeman-Grenville, G. S. P. (1962). The medieval history of the coast of Tanganyika. Oxford.

Garber, E. D. (1944). A cytological study of the genus *Sorghum*: sub-sections *Para-Sorghum* and *Eu-Sorghum*. Amer. Nat. 78: 89–94.

Gould, F. W. (1956). Chromosome counts and cytotaxonomic notes on grasses of the tribe *Andropogoneae*. Amer. Journ. Bot. 43: 395–404.

Grandbacher, F. J. (1963). The physiological function of the cereal awn. Bot. Rev. 29: 366–381.

Guilloteau, J. (1957). The problem of bush fires and burns in land development and soil conservation in Africa south of the Sahara. Sols Africains 4: 65–102.

Hackel, E. (1889). In De Candolle, Monographiae Phanerogamarum 6: 617–647.

Hitchcock, A. S. (1935). Manual of the grasses of the United States. Washington.

Hubbard, C. E. (1951). *Hyparrhenia edulis*. Hooker, Ic. Pl. 35, t. 3495.

Imperial Institute (1935). New materials for paper and board manufacture: grass (*Hyparrhenia sp.*) from Northern Rhodesia. Bull. Imp. Inst. 33: 428–430.

Jacques-Félix, H. (1950). Notes sur les Graminées d'Afrique tropicale. Rev. Bot. Appl. 30: 175–177.

—— (1962). Les Graminées d'Afrique tropicale. Paris.

Kemp, E. D. S., Mackenzie, R. M. & Romney, D. H. (1961). Productivity of pasture in British Honduras. III Jaragua grass. Trop. Agric., Trin. 38: 161–171.

Kinges, H. (1961). Merkmale des Grassembryos. Engler, Bot. Jahrb. 81: 50–93.

Krupko, S. (1953). Karyological studies and chromosome numbers in *Hyparrhenia aucta* and *H. hirta*. Journ. S. Afr. Bot. 19: 31–58.

—— (1955). Progress report on cytological study of *Hyparrhenia*. Ann. Rep. Frankenwald Field Res. Sta. 1954: 39.

—— (1956). *Hyparrhenia* chromosome studies. Ann. Rep. Frankenwald Field Res. Sta. 1955: 24–26.

Kuntze, O. (1891). Revisio generum plantarum 2: 789–793. Leipzig.

Larsen, K. (1954). Chromosome numbers of some European flowering plants. Bot. Tidsskr. 50: 163–174.

Linnaeus, C. (1737). Genera plantarum, ed. 1. Lugduni Batavorum.

Mansfeld, R. (1959). Vorläufiges Verzeichnis landwirtschaftlich oder gärtnerisch kultivierter Pflanzenarten. Kulturpfl., Beih. 2.

Mehra, K. L. (1955). Chromosome numbers in the tribe *Andropogoneae*. Ind. Journ. Genet. Pl. Breed. 15: 144.

—— & Anderson, E. (1960). Introgression between *Hyparrhenia cymbaria* Stapf and *H. papillipes* Anderss. in disturbed habitats of Ethiopia. Indian Journ. Genet. Pl. Breed. 20: 93–101.

Mes, M. C. (1952). The influence of some climatic factors on growth and seed production of grasses. South African Grassland Conf.: 39–51.

Metcalfe, C. R. (1960). The anatomy of the Monocotyledons I. Gramineae. Oxford.

Moffett, A. A. & Hurcombe, R. (1949). Chromosome numbers of South African grasses. Heredity 3: 369–373.

Moreau, R. E. (1952). Africa since the Mesozoic: with particular reference to certain biological problems. Proc. Zool. Soc. Lond. 121: 869–913.

—— (1963). Vicissitudes of the African biomes in the late Pleistocene. Proc. Zool. Soc. Lond. 141: 395–421.

Munro, P. E. (1966). Inhibition of nitrifiers by grass root extracts. Journ. Appl. Ecol. 3: 231–238.

Pilger, R. (1940). In Engler & Prantl, Die natürlichen Pflanzenfamilien 14e: 172–174. Leipzig.

—— (1954). Das System der *Gramineae* Engler, Bot. Jahrb. 76: 281–384.

Prat, H. (1932). L'épiderme des Graminées. Étude anatomique et systématique. Ann. Sci. Nat. Bot. sér. 10, 14: 117–324.

—— (1937). Charactères anatomiques et histologiques de quelques Andropogonées d'Afrique occidentale. Ann. Mus. Colon. Marseille sér. 5, 5, 2: 1–62.

Rattray, J. M. (1960). The grass cover of Africa. F.A.O. Agricultural Studies No. 49.

Reeder, J. R. (1957). The embryo in grass systematics. Amer. Journ. Bot. 44: 756–768.

Reichwaldt, E. (1945). Anatomische Untersuchungen an Andropogoneen Hüllspelzen. Biblioth. Bot. 120: 1–90.

Rendle, A. B. (1899). Catalogue of the African plants collected by Dr. Friedrich Welwitsch 2: 154–160. London.

Roberty, G. (1960). Monographie systématique des Andropogonées du globe. Boissiera 9: 1–455.

Schnell, R. (1961). Le problème des homologies phytogéographiques entre l'Afrique et l'Amérique tropicales. Mém. Mus. Hist. Nat. sér. B, 11: 137–241.

Schnell, R. (1962). Rémarques préliminaires sur quelques problèmes phytogéographiques du Sud-Est Asiatique. Rev. Gén. Bot. 69: 301–366.

Schweinfurth, G. (1867). Beitrag zur Flora Aethiopiens. Berlin.

Stapf, O. (1898). In Thistleton-Dyer, Flora Capensis 7: 334–366.

—— (1918). In Prain, Flora of Tropical Africa 9: 291–382.

Stebbins, G. L. (1956). Cytogenetics and evolution in the grass family. Amer. Journ. Bot. 43: 890–905.

—— (1957). Self-fertilization and population variability in the higher plants. Amer. Nat. 91: 337–354.

Stewart, D. R. M. (1965). The epidermal characters of grasses with special reference to East African species. Engler, Bot. Jahrb. 84: 63–116, 117–174.

Tateoka, T. (1965a). Chromosome numbers in some East African grasses. Amer. Journ. Bot. 52: 864–869.

—— (1965b). Contributions to biosystematic investigations of East African grasses. Bull. Nat. Sci. Mus. 8: 161–173.

—— (1965c). Chromosome numbers of some grasses from Madagascar. Bot. Mag., Tokyo 78: 306–311.

Tran, T. (1965). Les glumelles inférieures aristées de quelques graminées: anatomie, morphologie. Bull. Jard. Bot. Brux. 35: 219–284.

Vickery, J. W. (1935). The leaf anatomy and vegetative characters of the indigenous grasses of New South Wales. Proc. Linn. Soc. N.S. Wales 60: 340–373.

Vohra, S. K. (1966). Cytotaxonomy of some African grasses. M.Sc. thesis, London.

de Wet, J. M. J. (1954). Chromosome numbers of a few South African grasses. Cytologia 19: 97–103.

—— (1958). Additional chromosome numbers of Transvaal grasses. Cytologia 23: 113–118.

—— (1960). Chromosome numbers and some morphological attributes of various South African grasses. Amer. Journ. Bot. 47: 44–49.

—— & Anderson, L. J. (1956). Chromosome numbers in Transvaal grasses. Cytologia 21: 1–10.

White, F. (1965). The savanna woodlands of the Zambezian and Sudanian domains. Webbia 19: 651–681.

Zohary, D. & Feldman, M. (1962). Hybridization between amphidiploids and the evolution of polyploids in the wheat (Aegilops-Triticum) group. Evolution 16: 44–61.

INDEX TO COLLECTORS AND EXSICCATAE CITED

Bingham: 487 = 10; 606 = 7; 630 = 25

Blackburn: s.n. = 6a, 25

Blackwell in Eyles: 2233 = 42a

Blair: 2 = 7; 36 = 31; 45 = 21b; 49 = 3; 59 = 14; 65 = 8; 70 = 21a; 95 = 21a; 103 = 21b; 163 = 3; 174 = 3; 180 = 34; 181 = 34; 185 = 8; 204 = 3; 230 = 32; 242 = 32; 255 = 3; 350 = 43; 378 = 43

Blake: 9488 = 21b; 12801 = 6a

Blauchon: 16 = 6a; 53 = 8; 66 = 8

Boaler: 266 = 40; 537 = 52; 541 = 10; 599 = 15

Bogdan: 308 = 35; 407 = 35; 422 = 13; 427 = 30; 754 = 35; 761 = 34; 810 = 8; 1401 = 25; 1489 = 35; 1491 = 8; 1492 = 21b; 1547 = 3; 1634 = 8; 1663 = 31; 1910 = 27; 1911 = 27; 1912 = 13; 1914 = 35; 1952 = 6a; 1953 = 21b; 2069 = 30; 2162 = 24; 2181 = 3; 2564 = 6a; 2688 = 51; 2693 = 15; 2723 = 30; 2756 = 31; 2810 = 12; 2811 = 15; 2949 = 24; 3058 = 31; 3437 = 41; 3441 = 49; 3442 = 51; 3479 = 30; 3501 = 26; 3502 = 24; 3635 = 25; 3739 = 8; 3750 = 6a; 3752 = 21a; 4059 = 41; 4060 = 3; 4061 = 51; 4112 = 31; 4513 = 26; 4764 = 31; 5744 = 6a

Bogdan & Williams: 201 = 26

Boivin: 1983 = 28; s.n. = 6a, 21a

Bojer: s.n. = 24

Bolema: 993 = 21b

Bolus: 14986 = 12

Bonin: 406 = 51

Borgensen: 565 = 12

Bornmüller: 933 = 12

Bosser: 8002 = 21a; 10782 = 14; 10837 = 43a; 12474 = 25; 15992 = 42a; 16652 = 12; 17549 = 3; 18023 = 6a; 18171 = 28; 18317 = 6a; 18318 = 40

Boudet: 1550 = 6a; 1555 = 40; 1558 = 48b; 1896 = 39a; 2255 = 40; 3312 = 29; 3314 = 40

Boudet & Bille: 1446 = 19

Boughey in GC: 10752 = 4a; 12564 = 4a; 18628 = 4b

Bourgeau: 1504 = 12; 2812 = 12; 3270 = 41

Bouton: s.n. = 6a

Bové: s.n. = 12

Bovone: 50 = 9

Bowden: s.n. = 6a

Box: 131 = 6a

Brain: 3229 = 25; 3278 = 6a; 3531 = 41; 4239 = 35; 4242 = 26; 4249 = 42b; 10586 = 51; 10661 = 12

Brand in S.L.U.S.: 750 = 1

Brass: 16093 = 42b; 16275 = 24; 16975 = 41; 17090 = 7

Brazza & Thollon: 233 = 51

Bredo: 181 = 18; 2734 = 7; 2852 = 20; 2866 = 35; 4872 = 33

Brenan: 352 = 12

Breteler: 526 = 19; 528 = 40

Brockington: 27 = 35

Broun: 38 = 23

Brown: 20 = 34; 21 = 12; 28 = 51; 29 = 49

Brucckner: 373 = 12

Brunt: 591 = 33; 766 = 21a; 868 = 41; 870 = 6a; 875 = 51; 876 = 24; 882 = 39; 908 = 33; 977 = 35; 1057 = 19; 1126 = 19; 1129 = 6a; 1132 = 41; 1175 = 6a; 1176 = 33; 1178 = 42a; 1179 = 42a; 1182 = 6a; 1185 = 8; 1269 = 24; 1667 = 41

Buchanan: 37 = 24; 39 = 7; 223 = 21b; 228 = 24; 229 = 28; 303 = 7; 1423 = 3; s.n. = 8, 15, 35, 42a

Buechner: 6 = 8; 130 = 49

Buijsman: 52 = 25; 156 = 25

Bullock: 2955 = 50; 2958 = 21a; 2972 = 25; 3083 = 33; 3708 = 42b; 3949 = 6a

Bumpus: B6 = 34; Bam/26 = 41

Bunnemeyer: 11671 = 21b

Burchell: 4457 = 41

Burger: 933 = 3; 983 = 12; 1085 = 23; 2350 = 31; 2938 = 45

Burger & Getahun: 357 = 21b; 361 = 15

Burkart: 16814 = 12

Burnett: 49/110 = 23

Burton: s.n. = 6a

Burtt, B. D.: 1297 = 15; 1531 = 21a; 2037 = 35; 2570 = 23; 2597 = 24; 3953 = 24

Burtt-Davy: 469 = 13; 469a = 12; 1481 = 24; 1694 = 31; 2510/29 = 6a; 17211 = 12; 17807 = 28

Callens: 178 = 50; 1301 = 3; 1406 = 21a; 2484 = 42a; 3042 = 42a; 3588 = 51; 3591 = 39a; 3593 = 39a; 3608 = 42a; 3618 = 6a; 3805 = 11; 3893 = 11; 4052 = 19

Cartwright: 19 = 51

Chancellor: 211 = 25

Chand: 4650 = 14

Chandler: 1633 = 8; 1754 = 49

Chapman, E.: 692 = 12

Chapman, J. D.: 649 = 10; 651 = 25; 731 = 24

Charif: 1184 = 12

Charter in FHI: 36981 = 51; 36983 = 33

Chase: 6371 = 3

Chennels: 31 = 31

Chevalier: 2359 = 29; 4043 = 6a; 5313 = 39a; 5366 = 3; 5370 = 47; 5406 = 3; 5407 = 3; 5695 = 51; 5743 = 39a; 5925 = 18; 5948 = 18; 6829 = 6a; 7671 = 6a; 9795 = 17; 9850 = 17; 10406 = 6a; 10488 = 6a; 10490 bis = 16; 10507 = 17; 10509 = 5; 13785 = 51; 20091 = 51; 20185 = 50; 20196 = 40; 27285 = 19; 30664 = 51; 44806 = 12

Chillou: 1030 = 6a

Chiovenda: 1485 = 43; 1581 = 3; 1798 = 43; 1828 = 32; 2007 = 30; 2029 = 34; 2311 = 30; 2318 = 30; 2349 = 45; 2351 = 43; 2367 = 23; 2368 = 32; 2370 = 35; 2386 = 6a; 2428 = 31; 2460 = 34; 2480 = 3; 2536 = 35; 2601 = 28; 2606 = 34; 2609 = 32; 2626 = 32; 2702 = 28; 3116 = 24; 3221 = 10

Chipp: 5 = 25; 27 = 24

Claessens: 817 = 39; 1400 = 26

Clair: 8 = 39a

Clayton: 40 = 30; 85 = 20; 387 = 4b; 566 = 50; 567 = 4b; 568 = 29; 581 = 21b; 616 = 4b; 1379 =

17; 1434 = 41; 1448 = 21a; 1465 = 1; 1469 = 16; 4040 = 12; 4118 = 6a; 4134 = 6a; 4564 = 6a

Clemens: s.n. = 6a

Codd: 5963 = 7; 6880 = 21a; 7420 = 28

Collenette: 37 = 12

Colville: 85 = 23

Compère: 308 = 50; 1976 = 40; 2113 = 29

Compton: 25117 = 24; 26381 = 21b; 26427 = 21b; 26562 = 21b; 27775 = 24; 29946 = 21b; 30622 = 24

Corbett: 415 = 25

Corby: 4 = 3; 123 = 8; 385 = 42b; 387 = 15; GH 20978 = 41

Cormack: 3 = 10; 277 = 25; 279 = 24; 433 = 21a; 465 = 40; 475 = 6a

Craster: 74 = 21a; 59 = 43b

Crook: 421 = 3; 477 = 21a; 479 = 15; 484 = 42a; 501 = 9; 508 = 21b

Cruse: 331 = 42a

Czeczott: 201 = 12

Daggash in FHI: 24880 = 16

Dalziel: 263 = 16; 292 = 40; 293 = 4b; 295 = 18; 299 = 48a; 487 = 6a; 891 = 16; 901 = 50

Davey: 8 = 6a; FHI 27123 = 6a

Davies, J. H.: B35/63/OP = 6a

Davies, J. N.: 252 = 13; 257 = 50; 317 = 21a

Davies, R. M.: 78 = 21a; 361 = 25; 515 = 35; 520 = 10; 522 = 40; 1083 = 10; 1597a = 28; B26 = 26; B27 = 31; B28 = 28; B30 = 33; GH32694 = 13; GH 32696 = 13; GH 32697 = 41; GH 32698 = 34

Davis: 1696 = 12; 4183 = 12; 4625 = 12; 6247b = 12; 9561 = 12

Davis & Hedge: 26457 = 12

Dawe: 388 = 21a

Debeaux: s.n. = 12

Decary: 7329 = 42a; 14344 = 34; 15252 = 34

Deighton: 877 = 51; 3573 = 6a; 4205 = 6a; 4279 = 40; 4433 = 40; 4440 = 29; 4444 = 50; 4522 = 4b; 5448 = 6a

13

Richards: 1537 = 25; 9932 = 41; 12079 = 42a

Richardson: s.n. = 12

Riedel: 156 = 41

Ringoet: 3 = 25; 11 = 35 & 41

Risopoulos: 265 = 18; 268 = 50; 1234 = 50; 1237 = 6a; 1258 = 6a; 1259 = 48b; 1294 = 50; 1296 = 40; s.n. = 26

Roberts: 1745 = 13

Robinson, D. A.: 160 = 21b; 267 = 25; 351 = 24; 415 = 10

Robinson, E. A.: 185 = 7; 3080 = 2; 6481 = 24

Robson: 460 = 2

Robyns: 790 = 51; 1972 = 20

Rodin: 2888 = 12; 3624 = 12; 3821 = 15; 3822 = 31

Rogeon: 228 = 6a

Rogers, F. A.: 1327 = 7; 5945 = 15; 7701 = 21b; 10297 = 42b; 10399 = 42a; 11701 = 42a; 21124 = 9; 21169 = 12; 18575 = 24; s.n. = 12

Rose: s.n. = 6a

Rose Innes in GC: 30364 = 4b; 30448 = 48b; 30549 = 49; 30646 = 4b; 30676 = 48b; 30764 = 8; 30874 = 49; 31074 = 50; 31075 = 4b; 31099 = 1; 31171 = 50; 31172 = 6a; 31177 = 19; 31349 = 4b; 31391 = 18; 31469 = 6a; 31475 = 48b; 31480 = 50; 31509 = 50; 31514 = 4b; 31647 = 50; 32119 = 48b; 32353 = 6a; 32354 = 50; 32373 = 48b; 32385 = 4b; 32394 = 29; 32395 = 34; 32407 = 29; 32427 = 1; 32429 = 1; 32440 = 4b; 32441 = 29; 32470 = 50; 32471 = 29; 32472 = 48a; 32485 = 4b; 32487 = 4b; 32501 = 29; 32524 = 1; s.n. 12, 48b

Roseveare: 10/30a = 19; 16/30a = 51; 20/30a = 40

Rounsell: 3 = 6a

Rowainen: s.n. = 12

Ruddock: s.n. = 31

Ruxton: 32 = 29

Sacleux: 411 = 7

Sadio: 116 = 6a

de Saeger: 1471 = 19; 1472 = 8; 1491 = 21a; 1518 = 51; 1555 = 49

Sandwith, C.: 15 = 34; 99 = 12; s.n. = 28

Sandwith, N.Y.: 2576 = 12

Saunders: 9 = 41

Scaetta: 7 = 7; 28f = 7; 158 = 2; 1625 = 2; 1705 bis = 35; 2253 = 26; 3152 = 29

Schantz: 237 = 21a; 318 = 7; 900 = 21a; 994 = 3; 1061 = 24

Schantz & Turner: 4220 = 21a; 4228 = 19

Scheepers: 190 = 24; 215 = 6a; 242 = 34; 318 = 9; 1198 = 21a; 1203 = 21b

Schimper: 101 = 12; 138 = 35; 349 = 43; 408 = 28; 469 = 32; 584 = 33; 748 = 38c; 911 = 32; 923 = 23; 928 = 6a; 936 = 12; 937 = 32; 1006 = 31; 1009 = 27; 1010 = 45; 1033 = 30; 1052 = 28; 1055 = 30; 1056 = 12; 1089 = 23; 1116 = 33; 1118 = 6a; 1456 = 38a; 1458 = 37; 1637 = 43; 1797 = 10; 1805 = 1; 1821 = 45; 1822 = 23

Schlechter: 4054 = 42a

Schlieben: 732 = 12; 1007 = 42b; 1008 = 49; 1009 = 41; 3657 = 6a; 3815 = 7; 4633 = 27; 4634 = 35; 4635 = 3; 5093 = 12; 6263 = 20; 7201 = 9

Schmid: 2460 = 3

Schnell: 2358 = 21a; 3865 = 51

Schoenfelder: 72 = 12

Schultes & Black: 8489 = 6a

Schweickerdt: 1375 = 21a; 1772 = 32.

Schweinfurth: 482 = 12; 1034 = 38b; 1035 = 12; 1043 = 38c; 2421 = 20; 2602 = 51; 2618 = 39b; 2644 = 51

Schweinfurth & Riva: 1089 = 13; 2016 = 12; 2027 = 24

Scott: 70 = 31; 206 = 12

Seagrief: 3110 = 34

Seemann: 1562 = 41

Semsei in FH: 2161 = 7

Seret: 308 = 47

Shabani: s.n. = 27

5259 = 25; 5377 = 51; 5457b = 40; 5925 = 8; 6093 = 7; 7086 = 21b; 7381 = 21b

Torre & Paiva: 9170 = 3; 10677 = 3; 11273 = 40

Tothill: 2307 = 2

Townrow: 22 = 4b

Townsend: 63/95 = 12

Trapnell: 859 = 21b; 1446 = 42a; 1492 = 41; 1526 = 24; 1611 = 7; 1775 = 51; 2017 = 25; 2019 = 3; 2027 = 34; 2034 = 10; 2047 = 15; 2082 = 3

Trethewy: 175 = 12

Triana: 336 = 41

Troupin: 175 = 10; 826 = 51; 1695 = 10; 1904 = 10; 1905 = 25; 2007 = 5 & 6a; 3220 = 42a; 3587 = 25; 3589 = 7; 5671 = 27; 5784 = 21a; 5794 = 34; 5798 = 42b; 5861 = 42a; 5889 = 35; 5909 = 42b; 5921 = 42a; 6119 = 35; 6232 = 6a; 6523 = 21b; 6683 = 21a; 6741 = 35; 6836 = 34; 7188 = 21b; 7235 = 6a; 7595 = 24; 7598 = 21b; 9060 = 21a; 9578 = 21b; 9916 = 51

Tucker: 554 = 6a; 953 = 6a

Tuley: 60 = 6a; 489 = 50; 775 = 19; 933 = 18; 1023 = 41; 1030 = 19

Turnbull: 6 = 21a

Turrill: 1052 = 12

Unwin: 221 = 33

U Thein Lwin: 618 = 6b

Vaillant: 2148 = 51; 2291 = 39a; 2303 = 3; 2473 = 6a; 2477 = 16; 2478 = 34; 2548 = 48a; 2567 = 50; 2569 = 48a; 2570 = 50

Van der Ben: 656 = 24; 1534 = 6a; 1536 = 51; 1760 = 21b; 1816 = 21b; 1910 = 21b; 1969 = 51; 2009 = 19; 2032 = 51; 2045 = 21b; 2109 = 35; 2424 = 42a; 2458 = 21b; 2485 = 6a; 2525 = 6a

Vanderyst: 373 = 3; 414 = 42a; 3410 = 50; 3767 = 21a; 3773 = 6a; 3780 = 19; 3838 = 51; 4251

= 11; 5211 = 19; 6073 = 3; 15142 = 11; 23069 = 3; 27100 = 21b

Vasse: 209 = 34

Vaughan: A16 = 6a

Verboom: 542 = 21a; 702 = 34; 1366 = 15; 1370 = 34; 1385 = 25; 1394 = 6a

Verdcourt: 2619 = 8

Vesey-FitzGerald: 162b = 42a; 1648 = 3; 1654 = 25; 1664 = 50; 1710 = 34; 1712 = 51; 1742 = 40; 2272 = 3; 3381 = 42b; 5438 = 6a; 5630 = 6a; 15654/1 = 12; 15998/3 = 12; 17016/1 = 12; 17030/4 = 12; 17056/5 = 12; s.n. = 7

Viancin: s.n. = 39

Vickery: 12999 = 12; 13001 = 13

Vigne in FH: 3201 = 8; 4640 = 48b

Vogel: s.n. = 12

Volkens: 341 = 24

van Vuuren: 1360 = 15; 1684 = 15; 1685 = 10

Wager: C58 = 21b

Walker: s.n. = 51

Ward, C. J.: 2320 = 21a; 2548 = 9

Ward, J. F.: 39 = 29; L145 = 18

van Warmelo: 5157/4 = 24; 5159/16 = 21a

Warpur: 559 = 30

Watkins: 415 = 19; 422 = 42a

Weatherwax: 101 = 41

Welch: 81 = 20; 497 = 15

Welwitsch: 2728 = 22; 2820 = 39a; 2838 = 10; 3000 = 40; 7190 = 40; 7247 = 25; 7248 = 40; 7300 = 24; 7409 = 6a; 7515 = 42a; s.n. = 12

West: 249 = 32; 765 = 13

Wheeler-Haines: 370 = 51

Whellan: 1238 = 42b

White, F.: 7229 = 21a

White, O. E.: 1125 = 41

Whyte: s.n. = 42a, 51

Wickens: 656 = 38c; 979 = 3; 1059 = 23; 1108 = 6a; 1170 = 12; 1567 = 25; 2368 = 43; 2470 = 43; 2543 = 21b; 2991 = 15

Widgren: s.n. = 41

Wiehe (Nyasaland series): 39 = 27; 102 = 40; 107 = 21b, 180 = 31; 342 = 42b; 365 = 41; 371 = 6a;

INDEX TO NAMES OF TAXA

Names accepted in this revision are in bold type, synonyms are in italics, the remainder are excluded names.

Printed in England for Her Majesty's Stationery Office
by McCorquodale & Co Ltd, London, E.C.4
HM 3037 Dd. 142444 K6 9/69 McC 3309